Milk Products and Eggs

Fourth Supplement to McCance and Widdowson's

The Composition of Foods

FOR REFERENCE ONLY

Milk Products
and
Eggs

Fourth Supplement to

McCance and Widdowson's

The Composition of Foods

B. Holland, I. D. Unwin and D. H. Buss

The Royal Society of Chemistry
and
Ministry of Agriculture, Fisheries and Food

The Royal Society of Chemistry
Thomas Graham House
Science Park
Milton Road
Cambridge CB4 4WF
UK

Tel.: (0223) 4200 6 Telex: 818293

ISBN 0-85186-366-3

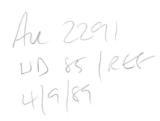

Orders should be addressed to:
The Royal Society of Chemistry
Distribution Centre
Letchworth
Herts. SG6 1HN
UK

Xerox Ventura Publisher™ output photocomposed and
printed in the United Kingdom by

Unwin Brothers Limited, Old Woking, Surrey

PREFACE

Following publication of the fourth edition of McCance and Widdowson's *The Composition of Foods* in 1978, the Ministry of Agriculture, Fisheries and Food took on the responsibility for maintaining and updating the official tables of food composition in the United Kingdom. In 1987 the Ministry joined with the Royal Society of Chemistry to begin production of a computerised UK National Nutrient Databank, from which this and future supplements to *The Composition of Foods* will be produced, as will an increasing variety of machine-readable products. This method of production will give greater accuracy, consistency and flexibility than is possible with manual compilation.

The availability of products from, and access to, the UK National Nutrient Databank are being developed simultaneously with the printed tables. As the nutrient values for each food group are updated and other data added, these are being merged with previously published data (principally from the fourth edition of *The Composition of Foods*) to bring the Databank progressively up-to-date. Thus nutrient values for the whole range of foods will be available for use with computer software, but only as material is revised will supporting information such as source references be added to the Databank. Some forms of access, for example through online search systems, may only become viable as the revision of food groups nears completion. The comments of users and potential users will influence the timing as well as the content and appearance of the products.

Epidemiological studies involving more than one country have heightened interest in international collaboration on the content of food composition tables. Much work has been done on food coding which seeks to describe food in detail and facilitate more precise computer retrieval, as well as to improve the comparability of nutrient data within and between countries. The RSC and MAFF will monitor and, where appropriate, participate in such developments so that the benefits can be available to users of the UK National Nutrient Databank.

CONTENTS

ACKNOWLEDGEMENTS

A large number of people have helped at each stage in the preparation of this book.

Most of the new analyses of milks, dairy products and eggs were performed at the Laboratory of the Government Chemist (LGC) and the Reading Laboratory of the AFRC Institute of Food Research (IFR, then the National Institute of Research in Dairying). The analytical teams were headed by Mr I Lumley and Mrs G Holcombe at the LGC and by Mr K J Scott and Mr E Florence at the IFR.

We are indebted to numerous manufacturers of dairy products and to manufacturers' organisations and retailers for information on the composition of their products. In particular we would like to thank Beecham Products, Birds Eye Wall's, Cambridge Nutrition, Cow & Gate, Express Foods Group, Farley Health Products, the Ice Cream Alliance, Mead Johnson Nutritionals, the Milk Marketing Board, Milupa, The Nestlé Company, Premier Brands UK, Raines Dairy Foods, St Ivel, Uni-Vite, Van den Berghs, Wander, and Wyeth Nutrition. The National Dairy Council kindly provided three of the cover photographs.

The final preparation of this book was overseen by a committee which, besides the authors, comprised of Miss P J Brereton (Northwick Park Hospital, Harrow), Dr J L Buttriss (National Dairy Council), Dr A M Fehily (MRC Epidemiology Unit, Cardiff), Miss A A Paul (MRC Dunn Nutritional Laboratory, Cambridge), Mr K J Scott (AFRC Institute of Food Research, Reading) and Professor D A T Southgate (AFRC Institute of Food Research, Norwich).

We would also like to express our appreciation for all the help given to us by so many people in the Ministry of Agriculture, Fisheries and Food, the Royal Society of Chemistry and elsewhere who were involved in the work leading to the production of this book.

INTRODUCTION

This is the fourth supplement to the fourth edition of McCance and Widdowson's *The Composition of Foods* (Paul and Southgate, 1978), and the second to replace part of that textbook. *The Composition of Foods* remains an essential text for those who need to know the nutritional value of foods consumed in Britain, but in the last ten years many new fresh and processed foods have become widely available, and the composition of many existing foods has changed. New analyses have resulted in more detailed information on many of these foods and nutrients. The Ministry of Agriculture, Fisheries and Food and the Royal Society of Chemistry are therefore collaborating in the development of a more comprehensive and up-to-date database on the nutrients in foods, information from which is being published in the present series of supplements. That which replaced the section on Cereals and Cereal Products was published in 1988 (Holland *et al.,* 1988), and the present supplement replaces and extends the sections on Milk and Milk Products and on Eggs.

Methods

A large number of new dairy and egg products were considered for inclusion, and even without the more transient items this supplement now contains 335 foods instead of the original 54. The selection of nutrient values has followed the general principles used in the preparation of the fourth edition - that is, from direct analyses wherever possible, together with appropriate values from the scientific literature and some carefully selected values from manufacturers. Many of these have been combined in recipes which make up about one third of the total number of foods.

Literature values

Nutrient values from the literature were taken from studies that included full details of the samples; where suitable methods of analysis had been used; and where the results had been presented in sufficient detail. They were from the UK wherever possible because the compositon of milk and its products may differ from that in countries with, for example, other breeds of cow, climates, and feeding and husbandry practices.

By analysis

Where a review of the literature (and, in some cases, of manufacturers' information) showed that little or nothing was known of the nutrients in an important food, arrangements were made for its direct analysis at the Laboratory of the Government Chemist or at the Reading Laboratory of the Institute of Food Research (Buss *et al.,* 1984; Florence *et al.,* 1984a, 1984b, 1985; Scott *et al.,* 1984a, 1984b; Scott and Bishop, 1986, 1988a, 1988b). Detailed sampling and analytical protocols were devised for each item. Most of the manufactured foods were bought from a wide variety of shops in London or Reading, with related items obtained as far as possible in proportion to their share of the market. Representative samples of pasteurised cows milk were, however, taken throughout the whole of Britain for 18 consecutive months. The samples were not normally analysed individually but, as for previous editions, pooled before analysis. The analytical methods were, in general, as described in *The Composition of Foods,* except for the following additions. Individual sugars were identified by thin layer chromatography and quantified by high performance liquid chromatography (HPLC) or enzymatically. Most minerals, including manganese, were determined by inductively coupled plasma emission spectroscopy, but selenium was determined by fluorimetry (Thorn *et al.*, 1978) and iodine by colorimetry (Wenlock *et al.*, 1982). Vitamin C was determined by microfluorimetry, and HPLC was also used in the determination of retinol isomers, vitamin D and vitamin E. Full details for each determination can be provided on request.

Although most of the nutrients in these dairy products and eggs were analysed, a few were derived by calculation. The fat soluble vitamins, fatty acids and cholesterol in many of the dairy products were calculated from the levels found in the milk fat from which they were derived, and the minerals and fat soluble vitamins in cooked eggs were derived from the amounts in raw eggs after allowing for any water losses.

Arrangement of the tables

Food groups

The dairy products and the eggs have each been arranged alphabetically within the following groups: liquid, concentrated and dried cows milks; other milks (including human milk); infant formulas; milk-based drinks; creams; cheeses; yogurts; ice creams; chilled and frozen dairy desserts and puddings; butter and selected fats for comparison; savoury milk- and cheese-based dishes and sauces; eggs; and egg-based dishes. This is broadly as in the fourth edition, but with a number of additional groups. A number of non-dairy analogues of dairy products have been included for easy cross-reference. A combined index and food coding list has been included at the end of this supplement to help in locating specific foods.

Numbering system

As in each edition of *The Composition of Foods* and in the supplement on cereal products, the foods have been renumbered in sequence starting from number 1. Dairy products continue until number 305, and then a gap has been left and the egg-based foods are numbered from 801 to 830. As in the previous supplement, however, a unique two digit prefix has been added to help distinguish between these and other broad groups of foods. For all dairy and egg products this code is '12', so that the full code number (and the one which will be used in Nutrient Databank applications) for, say, skimmed milk is 12-001 and for whole raw hens eggs is 12-801. A few foods which were also in the Cereals supplement had an additional number in the 11- series; they can be retrieved from the Databank using either number, but if their composition has changed in the meantime the higher number will give the most recent values.

Description and number of samples

The information given under this heading indicates the number and nature of the samples taken for analysis. The major sources of values that have been derived from the literature or from manufacturers' information, or by calculation, are also indicated under this heading. The manufacturer's name is given where this helps to identify a product. Where the calculation is from a recipe, the ingredients and the method of cooking are given at the end of this supplement (see also page 4).

Nutrients

The presentation of the nutrients has been changed slightly, not only from that in the fourth edition of *The Composition of Foods* but also from that in the previous supplement on *Cereals and Cereal Products*. This is to accommodate within four pages more information on fatty acids and individual sugars than before but without the detail on dietary fibre fractions that was desirable for cereals.

Proximates:–The first page for each food shows water, total nitrogen, protein, fat, available carbohydrate expressed as its monosaccharide equivalent, and energy value both in kilocalories and kilojoules. Protein was derived from the nitrogen values by multiplying them by the factors in **Table 1**, and the energy values were derived by multiplying the amounts of protein, fat and carbohydrate by the factors in **Table 2**. Energy from any lactic acid present (3.62 kcal or 15.1 kJ per g) was, however, not included. In human milk and in cocoa products where a substantial part of the nitrogen is not in protein, the true protein value is given.

Table 1   *Factors used for converting total nitrogen to protein*

Foods where all the protein is from milk	6.38
Foods where all the protein is from soya	5.70
All other foods, including soya-based infant formulas with added amino acids	6.25

Table 2   *Energy conversion factors*

	kcal/g	kJ/g
Protein	4	17
Fat	9	37
Available carbohydrate expressed as monosaccharide	3.75	16

Carbohydrates and fat:– The second page gives more details of the individual carbohydrates, fatty acids and cholesterol in the foods. The value for starch includes any modified starches, and that for total sugars is the sum of the values given for glucose, fructose, galactose, sucrose, maltose and lactose. For foods which contain glucose syrups, the sum of the starch and sugars will be less than the amount of carbohydrate shown on the preceding page; the difference consists of maltotriose and other oligosaccharides. Typical amounts of saturated fatty acids (including branched-chain acids), monounsaturated (*cis-* and *trans-* together) and polyunsaturated fatty acids are also shown, as well as the cholesterol content of the food. These values may, however, change in products made from vegetable fats if the blends of fat change. Full details of the individual fatty acids in foods, including those in dairy products and eggs, will be given in a future supplement.

Minerals and vitamins:– The minerals and vitamins are as in the third supplement on Cereals and Cereal Products, except that retinol and carotenes have been given separately because of their importance in dairy products. The retinol equivalent of a diet is conventionally taken as the sum of the retinol plus one-sixth of the β-carotene value. Where summer and winter values are given separately, summer is from May to October and winter from November to April. Recipe calculations use the average of these two values. A further table at the end of the book gives for a number of foods the amounts of all-*trans* retinol and 13-*cis* retinol, the latter having approximately 75 per cent of the biological activity of the *trans* form (Sivell *et al.*, 1984). Thiamin values are expressed as thiamin hydrochloride, vitamin B_6 as pyridoxine hydrochloride and pantothenate as calcium D-panthothenate. Folate values show total folic acid activity. For those soft and blue cheeses where the rind may be eaten, mineral and vitamin values are for the whole cheese.

Some loss of vitamins is inevitable when milk (or any other food) is stored. On the doorstep, milk exposed for several hours to bright sunlight can lose up to 70 per cent of its riboflavin. Vitamin C can also decline under those conditions from the 1 - 1.5 mg per 100 g in the original milk to almost zero. There will also be gradual losses of folate and vitamin B_{12} from UHT and sterilised milks even under ideal storage conditions because of reactions with small amounts of oxygen in the pack.

Additional tables

Finally, there are tables showing fibre fractions in those dairy and egg products which also contain cereal, and the amounts of vitamins claimed on the labels of some products which on analysis usually contain more (see main Tables) because of the need to ensure that enough of the labile vitamins remain to meet the claim even after storage.

All values are given per 100 grams of the edible portion of the food as described. However, because many dairy products may be sold or measured by volume, typical specific gravities (densities) of some of these products are given in **Table 3**.

Recipes   The recipes in this supplement were based on a variety of sources and may differ from those used in the fourth edition. The nutritional values assigned to the dairy products and eggs in each recipe were from this supplement, and for the cereals were from the third supplement. Wherever possible, updated information was used for other ingredients, too. Any weight losses were measured and assumed to consist only of water. Losses of labile vitamins were estimated for some recipes from the tables in the fourth edition, the relevant parts of which are reproduced in **Table 4**.

Other cooked eggs (boiled, fried with and without fat, poached, scrambled without milk) were analysed, and any differences between the values found for these and for raw eggs may be as much due to natural variability as to cooking losses.

Table 3   *Specific gravities of selected dairy products*

Skimmed milk	1.036
Semi-skimmed milk	1.034
Whole milk	1.031
Condensed milk (sweetened)	1.16
Evaporated milk (unsweetened)	1.066
Cambridge Diet (made up)	1.04
Single cream	1.00
Whipping cream	0.99
Double cream	0.99
Yogurts	1.08
	(range 1.03 - 1.2)
Ice cream	variable 0.5 - 0.6 approx
Eggs	1.02

Table 4   *Vitamins losses (%) on cooking, used in recipe calculations*

	Milk		Baked dishes	Eggs	
	Boiling[a]	Sauces[b]		Omelette	Scrambled
Thiamin	10	20	25	5	5
Riboflavin	10		15	20	20
Niacin			5	5	5
Vitamin B_6	10	20	25	15	15
Vitamin B_{12}	5				
Folate	20	50	50	30	30
Pantothenate	10	20	25	15	15
Vitamin C	50	50			
Vitamin E	20				

[a]   In milk-based drinks, custards, etc.
[b]   For example, for cheese sauce

Nutrient variability

Samples of the same or similar foods will always vary somewhat in composition. Thus, although we have given the values found on analysis of, for example, bulked samples of Cheddar cheeses from different countries, these differences may be as much due to sampling or analytical variations as to real differences between the foods. For this reason, manufacturers or retailers should analyse representative samples of each of their own products for nutrition labelling purposes. If the food or ingredient is sufficiently similar to one described in these tables, the value quoted herein may, however, serve as a guide.

A more comprehensive description of the factors to be taken into account both in the preparation and the proper use of food composition tables is given by Paul and Southgate in the introduction to *The Composition of Foods*. Users of the present supplement would be well advised to take them to heart.

References to Introductory text

Buss, D. H., Jackson, P. A. and Scuffam, D. (1984) Composition of butters on sale in Britain. *J. Dairy Res.* **51**, 637-641

Florence, E., Knight, D. J., Owen, J. A., Milner, D. F. and Harris W. M. (1985) Nutrient content of liquid milk as retailed in the United Kingdom. *J. Soc. Dairy Technol.* **38**, 121-127

Florence, E., Milner, D. F. and Harris, W. M. (1984a) Nutrient composition of dairy products. I. Cheeses. *J. Soc. Dairy Technol.* **37**, 13-16

Florence, E., Milner, D. F. and Harris, W. M. (1984b) Nutrient composition of dairy products. II. Creams. *J. Soc. Dairy Technol.* **37**, 16-18

Holland, B., Unwin, I.D., and Buss, D.H. (1988) *Third supplement to McCance and Widdowson's The Composition of Foods: Cereals and Cereal Products*, Royal Society of Chemistry, Letchworth

Paul, A. A. and Southgate, D. A. T. (1978) *McCance and Widdowson's The Composition of Foods Fourth edition*, HMSO, London

Scott, K. J. and Bishop, D. R. (1985) Nutrient content of milk and milk products : Water soluble vitamins in baby milk formulae. *J. Dairy Res.* **52**, 521-528

Scott, K. J. and Bishop, D. R. (1986) Nutrient content of milk and milk products : Vitamins of the B complex and vitamin C in retail market milk and milk products. *J. Soc. Dairy Technol.* **39**, 32-35

Scott, K. J. and Bishop, D. R. (1988a) Nutrient content of milk and milk products : Vitamins of the B complex and vitamin C in retail cheeses. *J. Sci. Food Agric.* **43**, 187-192

Scott, K. J., and Bishop, D. R. (1988b) Nutrient content of milk and milk products : Vitamins of the B complex and vitamin C in retail creams, ice creams and milk shakes. *J. Sci. Food Agric.* **43**, 193-199

Scott, K. J., Bishop, D. R., Zechalko, A., Edwards-Webb, J. D., Jackson, P. A., and Scuffam, D. (1984a) Nutrient content of liquid milk. I. Vitamins A, D_3, C and of the B complex in pasteurized bulk liquid milk. *J. Dairy Res.* **51**, 37-50

Scott, K. J., Bishop, D. R., Zechalko, A. and Edwards-Webb, J. D. (1984b) Nutrient content of liquid milk. II. Content of vitamin C, riboflavin, folic acid, thiamin, vitamins B_{12} and B_6 in pasteurized milk as delivered to the home and after storage in the domestic refrigerator. *J. Dairy Res.* **51**, 51-57

Sivell, L. M., Bull, N. L., Buss, D. H., Wiggins, R. A., Scuffam, D., and Jackson, P. A. (1984) Vitamin A activity in foods of animal origin. *J. Sci. Food Agric.* **35**, 931-939

Thorn, J., Robertson, J., and Buss, D. H. (1978) Trace nutrients. Selenium in British food. *Br. J. Nutr.* **39**, 391-396

Wenlock, R. W., Buss, D. H., Moxon, R. E., and Bunton, N. G. (1982) Trace nutrients 4. Iodine in British food. *Br. J. Nutr.* **47**, 381-390

Symbols and abbreviations used in the tables

Symbols

0	None of the nutrient is present
Tr	Trace
N	The nutrient is present in significant quantities but there is no reliable information on the amount
( )	Estimated value
Italic text	Carbohydrate or starch estimated 'by difference', and energy values based upon these quantities

Abbreviations

Gluc	Glucose
Fruct	Fructose
Galact	Galactose
Sucr	Sucrose
Malt	Maltose
Lact	Lactose
Satd	Saturated fatty acids
Monounsatd	Monounsaturated fatty acids
Polyunsatd	Polyunsaturated fatty acids
Trypt	Tryptophan
SMP	Dried Skimmed Milk Powder
UHT	Ultra High Temperature heat sterilisation process

Milks

and

Milk Products

Cows milks

Composition of food per 100g

No. 12-	Food	Description and main data sources	Water g	Total nitrogen g	Protein g	Fat g	Carbo-hydrate g	Energy value kcal	kJ
1	**Skimmed milk**, *average*	Weighted average of pasteurised, sterilised and UHT	91.1	0.52	3.3	0.1	5.0	33	140
2	*pasteurised*	10 samples	91.1	0.52	3.3	0.1	5.0	33	140
3	*fortified plus SMP*	10 samples, own label and Vitapint	89.3	0.60	3.8	0.1	6.0	39	164
4	*UHT*	10 samples	91.1	0.53	3.4	0.1	4.9	32	137
5	*fortified*	9 samples	90.9	0.54	3.5	0.2	5.0	35	147
6	*fortified plus SMP*	10 samples, Slimcea	89.9	0.59	3.8	0.3	5.5	38	162
7	*sterilised*	10 samples, 2 brands, polybottles	90.9	0.54	3.5	0.1	4.8	32	138
8	**Semi-skimmed milk**, *average*	Weighted average of pasteurised and UHT	89.8	0.52	3.3	1.6	5.0	46	195
9	*pasteurised*	10 samples	89.8	0.52	3.3	1.6	5.0	46	195
10	*fortified plus SMP*	10 samples, own label and Vitapint	88.4	0.59	3.7	1.6	5.8	51	215
11	*UHT*	10 samples	89.7	0.52	3.3	1.7	4.8	46	194
12	**Whole milk**, *average*	Weighted average of pasteurised, sterilised and UHT	87.8	0.50	3.2	3.9	4.8	66	275
13	*pasteurised*[a]	186 samples, bottles and cartons. Fat from Milk Marketing Board	87.8	0.50	3.2	3.9	4.8	66	275
14	*summer*	Selected nutrients only	87.8	0.50	3.2	3.9	4.8	66	275
15	*winter*	Selected nutrients only	87.8	0.50	3.2	3.9	4.8	66	275
16	*UHT*	10 samples	87.8	0.50	3.2	3.9	4.8	66	275
17	*sterilised*	10 samples, 2 brands, polybottles	87.6	0.55	3.5	3.9	4.5	66	277

a All the values for pasteurised milk are equally applicable to unpasteurised milk

12

Cows milks

Carbohydrate fractions and fatty acids, g per 100g
Cholesterol, mg per 100g

No. 12-	Food	Starch	Total sugars	Individual sugars						Fatty acids			Cholest-erol
				Gluc	Fruct	Galact	Sucr	Malt	Lact	Satd	Mono unsatd	Poly unsatd	
1	**Skimmed milk**, *average*	0	5.0	0	0	0	0	0	5.0	0.1	Tr	Tr	2
2	pasteurised	0	5.0	0	0	0	0	0	5.0	0.1	Tr	Tr	2
3	*fortified plus SMP*	0	6.0	0	0	0	0	0	6.0	0.1	Tr	Tr	2
4	UHT	0	4.9	0	0	0	0	0	4.9	0.1	Tr	Tr	2
5	*fortified*	0	5.0	0	0	0	0	0	5.0	0.1	0.1	Tr	2
6	*fortified plus SMP*	0	5.5	0	0	0	0	0	5.5	0.2	0.1	Tr	2
7	sterilised	0	4.8	0	0	0	0	0	4.8	0.1	Tr	Tr	2
8	**Semi-skimmed milk**, *average*	0	5.0	0	0	0	0	0	5.0	1.0	0.5	Tr	7
9	pasteurised	0	5.0	0	0	0	0	0	5.0	1.0	0.5	Tr	7
10	*fortified plus SMP*	0	5.8	0	0	0	0	0	5.8	1.0	0.5	Tr	7
11	UHT	0	4.8	0	0	0	0	0	4.8	1.1	0.5	Tr	7
12	**Whole milk**, *average*	0	4.8	0	0	0	0	0	4.8	2.4	1.1	0.1	14
13	pasteurised	0	4.8	0	0	0	0	0	4.8	2.4	1.1	0.1	14
14	*summer*	0	4.8	0	0	0	0	0	4.8	2.4	1.2	0.1	14
15	*winter*	0	4.8	0	0	0	0	0	4.8	2.5	1.1	0.1	14
16	UHT	0	4.8	0	0	0	0	0	4.8	2.4	1.1	0.1	14
17	sterilised	0	4.5	0	0	0	0	0	4.5	2.4	1.1	0.1	14

Cows milks

Inorganic constituents per 100g

No. 12-	Food	Na	K	Ca	Mg	P	Fe	Cu	Zn	S	Cl	Mn	Se	I
							mg						µg	
1	**Skimmed milk**, *average*	54	150	120	12	94	0.06	Tr	0.4	31	100	Tr	(1)	(15)
2	pasteurised	55	150	120	12	95	0.05	Tr	0.4	31	100	Tr	(1)	(15)
3	*fortified plus SMP*	61	170	140	13	110	0.04	Tr	0.4	36	110	Tr	(1)	(15)
4	UHT	51	150	110	11	89	0.10	Tr	0.4	31	100	Tr	(1)	(15)
5	*fortified*	54	150	110	10	89	0.08	Tr	0.3	31	100	Tr	(1)	(15)
6	*fortified plus SMP*	60	160	120	11	100	0.07	Tr	0.4	36	110	Tr	N	N
7	sterilised	51	140	120	12	95	0.09	0.01	0.4	31	100	Tr	(1)	(15)
8	**Semi-skimmed milk**, *average*	55	150	120	11	95	0.05	Tr	0.4	31	100	Tr	(1)	(15)
9	pasteurised	55	150	120	11	95	0.05	Tr	0.4	31	100	Tr	(1)	(15)
10	*fortified plus SMP*	59	150	130	12	100	0.03	Tr	0.4	35	110	Tr	(1)	(15)
11	UHT	50	150	110	11	90	0.17	Tr	0.4	31	100	Tr	(1)	(15)
12	**Whole milk**, *average*	55	140	115	11	92	0.06	Tr	0.4	30	100	Tr	1	15
13	pasteurised	55	140	115	11	92	0.05	Tr	0.4	30	100	Tr	1	15
14	*summer*	55	140	115	11	92	0.05	Tr	0.4	30	100	Tr	1	7
15	*winter*	55	140	115	11	92	0.05	Tr	0.4	30	100	Tr	1	37
16	UHT	51	140	110	11	87	0.23	0.01	0.4	30	93	Tr	(1)	(15)
17	sterilised	57	140	120	13	91	0.18	Tr	0.3	30	100	Tr	(1)	(15)

Cows milks

No. 12-	Food	Retinol µg	Carotene µg	Vitamin D µg	Vitamin E mg	Thiamin mg	Ribo- flavin mg	Niacin mg	Trypt 60 mg	Vitamin B6 mg	Vitamin B12 µg	Folate µg	Panto- thenate mg	Biotin µg	Vitamin C mg
1	**Skimmed milk**, *average*	1	Tr	Tr	Tr	0.04	0.17	0.09	0.78	0.06	0.4	5	0.32	1.9	1
2	pasteurised	1	Tr	Tr	Tr	0.04	0.18	0.09	0.78	0.06	0.4	6	0.32	2.0	1
3	*fortified plus SMP*	43	5	0.26	0.01	0.04	0.19	0.12	0.90	0.06	0.4	5	0.40	2.4	1
4	*UHT*	1	Tr	Tr	Tr	0.04	0.17	0.09	0.79	0.05	0.2	1	0.32	1.7	Tr
5	*fortified*	61	18	0.07	0.02	0.04	0.18	0.10	0.81	0.05	Tr	4	0.33	1.5	35[a]
6	*fortified plus SMP*	50	5	0.08	0.02	0.04	0.16	0.11	0.89	0.05	Tr	5	0.38	1.5	38[a]
7	*sterilised*	1	Tr	Tr	Tr	0.03	0.14	0.10	0.81	0.04	0.1	Tr	0.32	2.0	Tr
8	**Semi-skimmed milk**, *average*	21	9	0.01	0.03	0.04	0.18	0.09	0.78	0.06	0.4	6	0.32	2.0	1
9	pasteurised	21	9	0.01	0.03	0.04	0.18	0.09	0.78	0.06	0.4	6	0.32	2.0	1
10	*fortified plus SMP*	90	5	0.13	0.04	0.04	0.19	0.10	0.89	0.06	0.4	5	0.37	2.3	1
11	*UHT*	20	11	0.01	0.03	0.04	0.18	0.09	0.78	0.05	0.2	2	0.33	1.8	Tr
12	**Whole milk**, *average*	52	21	0.03	0.09	0.03	0.17	0.08	0.75	0.06	0.4	6	0.35	1.9	1
13	pasteurised	52	21	0.03	0.09	0.04	0.17	0.08	0.75	0.06	0.4	6	0.35	1.9	1
14	*summer*	62	31	0.03	0.10	0.04	0.17	0.08	0.75	0.06	0.4	4	0.35	1.9	1
15	*winter*	41	11	0.03	0.07	0.04	0.17	0.08	0.75	0.06	0.4	7	0.35	1.9	1
16	*UHT*	47	25	0.03	0.07	0.04	0.18	0.09	0.75	0.04	0.2	1	0.32	1.8	Tr
17	*sterilised*	52	21	0.03	0.09	0.03	0.14	0.09	0.83	0.04	0.1	Tr	0.28	1.8	Tr

[a] Unfortified milks would contain only traces of vitamin C

No. 12-	Food	Description and main data sources	Water g	Total nitrogen g	Protein g	Fat g	Carbo-hydrate g	Energy value kcal	kJ
18	**Channel Island milk,** whole, pasteurised	Samples from dairy and retail outlets. Fat from Milk Marketing Board	86.4	0.57	3.6	5.1	4.8	78	327
19	*summer*	Selected nutrients only	86.4	0.57	3.6	5.1	4.8	78	327
20	*winter*	Selected nutrients only	86.4	0.57	3.6	5.1	4.8	78	327
21	semi-skimmed, UHT	10 samples	89.4	0.57	3.6	1.6	4.8	47	197
22	**Buttermilk**	3 cultured samples (Eden Vale)	90.4	0.53	3.4	0.5	5.0	37	157
23	**Buttermilk powder**	Commercial ingredient, calculated from uncultured buttermilk	3.0	5.35	34.1	5.1	51.0	373	1584
24	**Calcium-fortified milk,** Calcia	8 samples	87.6	0.62	4.0	0.5	6.2	44	186
25	Vital	8 samples	88.7	0.62	4.0	0.1	5.2	36	155
26	**Coffee Compliment**	Manufacturer's data (Premier Brands)	3.5	0.47	3.0	36.0	*58.1*	*554*	*2313*
27	**Coffeemate**	Analysis and manufacturer's data (Carnation)	3.0	0.42	2.7	34.9	*57.3[a]*	*540*	*2254*
28	**Condensed milk,** skimmed, *sweetened*	10 cans (Fussells)	29.7	1.57	10.0	0.2	60.0	267	1137
29	whole, *sweetened*	10 cans, 2 brands	25.9	1.33	8.5	10.1	55.5	333	1406
30	**Dried skimmed milk**	20 samples, 7 brands, fortified	3.0	5.70	36.1	0.6	52.9	348	1482
31	*with vegetable fat*	12 samples, 5 brands, fortified	2.0	3.70	23.3	25.9	42.6	487	2038
32	**Dried whole milk**	Mixed sample, different brands. Vitamins calculated	2.9	4.12	26.3	26.3	39.4	490	2051

[a] Including oligosaccharides from the glucose syrup/maltodextrins in the product

Carbohydrate fractions and fatty acids, g per 100g
Cholesterol, mg per 100g

No. 12-	Food	Starch	Total sugars	Individual sugars						Fatty acids			Cholest-erol
				Gluc	Fruct	Galact	Sucr	Malt	Lact	Satd	Mono unsatd	Poly unsatd	
18	**Channel Island milk,**												
	whole, pasteurised	0	4.8	0	0	0	0	0	4.8	3.3	1.3	0.1	16
19	*summer*	0	4.8	0	0	0	0	0	4.8	3.2	1.4	0.1	16
20	*winter*	0	4.8	0	0	0	0	0	4.8	3.3	1.3	0.1	16
21	semi-skimmed, UHT	0	4.8	0	0	0	0	0	4.8	1.0	0.4	Tr	7
22	**Buttermilk**	0	5.0	0	0	0	0	0	5.0	0.3	0.1	Tr	2
23	**Buttermilk powder**	0	51.0	0	0	0	0	0	51.0	3.2	1.5	0.1	20
24	**Calcium-fortified milk,**												
	Calcia	0	6.2	0	0	0	0	0	6.2	0.3	0.1	Tr	(6)
25	Vital	0	5.2	0	0	0	0	0	5.2	0.1	Tr	Tr	2
26	**Coffee Compliment**	Tr	N	N	0	0	0	N	0	35.4	N	N	Tr
27	**Coffeemate**	Tr	9.8[a]	5.2	0	0	0	4.6	Tr	32.1	1.1	Tr	2
28	**Condensed milk,**												
	skimmed, sweetened	0	60.0	0	0	0	46.7	0	13.3	0.1	0.1	Tr	1
29	*whole, sweetened*	0	55.5	0	0	0	43.2	0	12.3	6.3	2.9	0.3	36
30	**Dried skimmed milk**	0	52.9	0	0	0	0	0	52.9	0.4	0.2	Tr	12
31	*with vegetable fat*	0	42.6	0	0	0	0	0	42.6	16.8	7.3	0.7	17
32	**Dried whole milk**	0	39.4	0	0	0	0	0	39.4	16.5	7.6	0.8	120

a Not including oligosaccharides from the glucose syrup/maltodextrins in the product

Cows milks continued

Inorganic constituents per 100g

No. 12-	Food	Na	K	Ca	Mg	P	Fe	Cu	Zn	S	Cl	Mn	Se (µg)	I (µg)
							mg							
18	**Channel Island milk,** whole, pasteurised	54	140	130	12	100	0.05	Tr	0.4	30	100	Tr	(1)	N
19	summer	54	140	130	12	100	0.06	Tr	0.4	30	100	Tr	(1)	N
20	winter	54	140	130	12	100	0.06	Tr	0.4	30	100	Tr	(1)	N
21	semi-skimmed, UHT	55	140	120	11	100	0.14	Tr	0.4	30	97	Tr	N	N
22	**Buttermilk**	56	150	120	13	95	0.02	Tr	0.4	32	110	Tr	(1)	N
23	**Buttermilk powder**	570	1510	1210	130	960	0.20	Tr	3.7	320	1110	Tr	(11)	N
24	**Calcium-fortified milk,** Calcia	55	190	170	13	110	(0.05)	Tr	0.4	37	N	Tr	(1)	N
25	Vital	51	190	180	12	110	(0.05)	Tr	0.2	37	130	Tr	(1)	N
26	**Coffee Compliment**	800	10	15	N	N	N	N	0.2	N	N	N	N	N
27	**Coffeemate**	200	900	4	N	350	N	N	N	N	N	N	N	N
28	**Condensed milk,** skimmed, sweetened	150	450	330	33	270	0.33	Tr	1.2	94	300	Tr	(3)	(89)
29	whole, sweetened	140	360	290	29	240	0.23	Tr	1.0	81	230	Tr	(3)	74
30	**Dried skimmed milk**	550	1590	1280	130	970	0.27	Tr	4.0	320	1070	Tr	(11)	(150)
31	with vegetable fat	440	1030	840	74	680	0.19	Tr	0.6	220	760	Tr	(7)	N
32	**Dried whole milk**	440	1270	1020	84	740	0.40	0.02	3.2	240	810	Tr	(8)	(120)

18

Cows milks *continued*

No. 12-	Food	Retinol µg	Carotene µg	Vitamin D µg	Vitamin E mg	Thiamin mg	Ribo-flavin mg	Niacin mg	Trypt 60 mg	Vitamin B6 mg	Vitamin B12 µg	Folate µg	Panto-thenate mg	Biotin µg	Vitamin C mg
18	**Channel Island milk,**														
	whole, pasteurised	46	71	0.03	0.11	0.04	0.19	0.07	0.85	0.06	0.4	6	0.36	1.9	1
19	*summer*	65	115	0.04	0.13	0.04	0.19	0.07	0.85	0.06	0.4	5	0.36	1.9	1
20	*winter*	27	27	0.03	0.09	0.04	0.19	0.07	0.85	0.06	0.4	7	0.36	1.9	1
21	semi-skimmed, UHT	14	22	0.01	0.04	0.04	0.19	0.09	0.85	0.05	0.2	1	0.34	1.5	Tr
22	**Buttermilk**	7	3	Tr	0.01	0.04	0.18	0.10	0.79	0.05	0.1	9	0.32	2.1	(1)
23	**Buttermilk powder**	71	30	0.04	0.12	0.36	1.82	1.03	8.03	0.51	0.8	N	3.23	19.1	N
24	**Calcium-fortified milk,**														
	Calcia	6	Tr	0.19	0.03	0.06	0.52	0.03	0.93	0.05	1.8	4	N	N	N
25	Vital	Tr	Tr	0.26	0.03	0.31	0.36	0.06	0.93	0.05	2.6	4	N	N	N
26	**Coffee Compliment**	0	Tr	0	N	0	0	0	0.71	0	0	0	0	0	0
27	**Coffeemate**	0	200	0	N	0	1.00	0	0.63	0	0	0	0	0	0
28	**Condensed milk,**														
	skimmed, sweetened	28	20	0.85	0.04	0.11	0.51	0.30	2.35	0.09	0.9	16	1.03	5.2	5
29	whole, sweetened	110	70	5.40	0.19	0.09	0.46	0.29	1.99	0.07	0.7	15	0.85	3.9	4
30	**Dried skimmed milk** [a]	350	5	2.10	0.27	0.38	1.63	1.02	8.55	0.60	2.6	51	3.28	20.1	13
31	*with vegetable fat*	395	15	10.50	1.32	0.23	1.20	0.63	5.55	0.35	2.3	36	2.15	15.0	11
32	**Dried whole milk**	290	170	0.24	0.61	0.31	1.40	0.60	6.18	0.48	2.4	46	2.79	13.9	9

[a] Unfortified skimmed milk powder contains approximately 8µg retinol, 3µg carotene, Tr vitamin D, and 0.01mg vitamin E per 100g. Some brands contain as much as 755µg retinol, 10µg carotene and 4.6µg vitamin D per 100g

Cows milks *continued*

Composition of food per 100g

No. 12-	Food	Description and main data sources	Water g	Total nitrogen g	Protein g	Fat g	Carbo-hydrate g	Energy value kcal	Energy value kJ
33	**Evaporated milk**, whole[a]	12 samples, Ideal, Carnation and own brands	69.1	1.32	8.4	9.4	8.5	151	629
34	**Flavoured milk**	32 samples in polybottles; mixed flavours, sterilised; skimmed and whole milk	85.4	0.56	3.6	1.5	10.6[b]	68	287
35	– **Whey**	Commercial ingredient, sweet whey, literature sources	93.3	0.16	1.0	0.2	5.1	25	106
36	dried	Commercial ingredient, sweet whey, literature sources	4.0	1.91	12.2	1.3	78.0	353	1503

[a] Carnation Light (reduced fat evaporated milk) contains 9.2g protein, 4.0g fat, 13.1g carbohydrate, 122 kcals and 514 kJ per 100g

[b] Including oligosaccharides from the glucose syrup/maltodextrins in the product

Cows milks *continued*

Carbohydrate fractions and fatty acids, g per 100g
Cholesterol, mg per 100g

No. 12-	Food	Starch	Total sugars	Individual sugars						Fatty acids			Cholest-erol
				Gluc	Fruct	Galact	Sucr	Malt	Lact	Satd	Mono unsatd	Poly unsatd	
33	**Evaporated milk**, whole	0	8.5	0	0	0	0	0	8.5	5.9	2.7	0.3	34
34	**Flavoured milk**	Tr	9.4[a]	Tr	0	0	4.3	Tr	5.1	0.9	0.4	Tr	7
35	**Whey**	0	5.1	0	0	0	0	0	5.1	0.1	0.1	Tr	Tr
36	dried	0	78.0	0	0	0	0	0	78.0	0.8	0.4	Tr	5

[a] Not including oligosaccharides from the glucose syrup/maltodextrins in the product

Cows milks *continued*

No. 12-	Food	Na	K	Ca	Mg	P	mg Fe	Cu	Zn	S	Cl	Mn	µg Se	I
33	**Evaporated milk**, whole	180	360	290	29	260	0.26	0.02	0.9	84	250	Tr	(3)	11
34	**Flavoured milk**	61	150	110	13	89	0.23	Tr	0.5	N	110	Tr	N	N
35	**Whey**	50	140	55	8	50	0.08	Tr	0.1	N	95	N	N	N
36	dried	1090	1980	790	150	770	0.90	N	1.2	N	1500	N	N	N

Cows milks *continued*

No. 12-	Food	Retinol μg	Carotene μg	Vitamin D μg	Vitamin E mg	Thiamin mg	Ribo-flavin mg	Niacin mg	Trypt 60 mg	Vitamin B6 mg	Vitamin B12 μg	Folate μg	Panto-thenate mg	Biotin μg	Vitamin C mg
33	Evaporated milk, whole	105	100	3.95[a]	0.19	0.07	0.42	0.23	1.98	0.07	0.1	11	0.75	4.0	1
34	Flavoured milk	20	8	0.01	0.03	0.03	0.17	0.11	0.84	0.03	0.1	2	0.30	2.2	Tr
35	Whey	4	Tr	Tr	Tr	0.04	0.13	0.07	0.24	0.05	0.2	4	0.36	1.4	1
36	dried	15	Tr	Tr	Tr	0.50	2.40	0.95	2.87	0.50	2.4	22	5.60	4.3	5

[a] This is for fortified product. Unfortified evaporated milk contains approximately 0.09μg vitamin D per 100g

Other milks

Composition of food per 100g

No. 12-	Food	Description and main data sources	Water g	Total nitrogen g	Protein g	Fat g	Carbo-hydrate g	Energy value kcal	kJ
37	**Goats milk**, pasteurised	20 samples from one herd and literature sources	88.9	0.49	3.1	3.5	4.4	60	253
38	**Human milk**, colostrum	Literature sources	88.2	0.31	2.0	2.6	6.6	56	236
39	transitional	Mixed sample, 15 mothers at 10th day post partum and literature sources	(87.4)	0.23	1.5	3.7	6.9	67	281
40	mature	DHSS (ref. 1) and literature sources	87.1	0.20	1.3[a]	4.1	7.2	69	289
41	**Sheeps milk**, *raw*	30 samples from 2 herds and literature sources	83.0	0.85	5.4	6.0	5.1	95	396
42	**Soya milk**, plain	6 samples, 4 brands	89.7	0.52	2.9	1.9	0.8	32	132
43	flavoured	4 brands, assorted flavours	89.4	0.49	2.8	1.7	3.6	40	168

[a] $N \times 6.38$. Excluding the non-protein nitrogen, true protein = 0.85g per 100g

Carbohydrate fractions and fatty acids, g per 100g
Cholesterol, mg per 100g

No. 12-	Food	Starch	Total sugars	Individual sugars						Fatty acids			Cholest- erol
				Gluc	Fruct	Galact	Sucr	Malt	Lact	Satd	Mono unsatd	Poly unsatd	
37	**Goats milk**, pasteurised	0	4.4	0	0	0	0	0	4.4	2.3	0.8	0.1	10
38	**Human milk**, colostrum	0	6.6	0	0	0	0	0	6.6	1.1	1.1	0.3	31
39	transitional	0	6.9	0	0	0	0	0	6.9	1.5	1.5	0.5	24
40	mature	0	7.2	0	0	0	0	0	7.2	1.8	1.6	0.5	16
41	**Sheeps milk**, *raw*	0	5.1	0	0	0	0	0	5.1	3.8	1.5	0.3	11
42	**Soya milk**, plain	0	0.8	0	0	0	0.8	0	0	0.3	0.4	1.1	0
43	flavoured	0	3.6	0.3	0.2	0	3.1	0	0	(0.2)	(0.4)	(1.0)	0

Other milks

Inorganic constituents per 100g

No. 12-	Food	Na	K	Ca	Mg	P	Fe	Cu	Zn	S	Cl	Mn	Se	I
							mg						µg	
37	**Goats milk**, pasteurised	42	170	100	13	90	0.12	0.03	0.5	N	150	Tr	N	N
38	**Human milk**, colostrum	47	70	28	3	14	0.07	0.05	0.6	22	N	Tr	N	N
39	transitional	30	57	25	3	16	0.07	0.04	(0.3)	19	86	Tr	(2)	N
40	mature	15	58	34	3	15	0.07	0.04	0.3	14	42	Tr	1	7
41	**Sheeps milk**, *raw*	44	120	170	18	150	0.03	0.10	0.7	N	82	Tr	N	N
42	**Soya milk**, plain	32	120	13	15	47	0.40	0.06	0.2	N	15	0.1	N	N
43	flavoured	61	110	14	18	53	0.40	0.02	0.2	N	130	0.1	N	N

No. 12-	Food	Retinol µg	Carotene µg	Vitamin D µg	Vitamin E mg	Thiamin mg	Ribo-flavin mg	Niacin mg	Trypt 60 mg	Vitamin B6 mg	Vitamin B12 µg	Folate µg	Panto-thenate mg	Biotin µg	Vitamin C mg
37	Goats milk, pasteurised	44	Tr	0.11	0.03	0.04	0.13	0.31	0.73	0.06	0.1	1	0.41	3.0	1
38	Human milk, colostrum	155	(135)	N	1.30	Tr	0.03	0.05	0.72	Tr	0.1	2	0.12	Tr	7
39	transitional	85	(37)	N	0.48	0.01	0.03	0.14	0.54	Tr	0.03	3	0.20	0.2	6
40	mature	58	(24)	0.04	0.34	0.02	0.03	0.22	0.47	0.01	0.01	5	0.25	0.7	4
41	Sheeps milk, raw	83	Tr	0.18	0.11	0.08	0.32	0.41	1.27	0.08	0.6	5	0.45	2.5	5.
42	Soya milk, plain	0	Tr	0[a]	0.74	0.06	0.27	0.11	0.52	0.07	0[a]	19	N	N	0
43	flavoured	0	N	0[a]	0.66	0.06	0.03	0.11	0.49	0.07	0[a]	15	N	N	Tr

[a] Some brands contain added vitamins D and B_{12}

Infant formulas

12-044 to 12-059
Composition of food per 100g

No. 12-	Food	Description and main data sources	Water g	Total nitrogen g	Protein g	Fat g	Carbo-hydrate g	Energy value kcal	kJ
	Whey-based modified milks								
44	**Aptamil**	Analysis and manufacturer's data (Milupa)	3.2	1.99	12.7	29.3	56.4	526	2190
45	*reconstituted*	Calcd. as 1 scoop granules (4.33g) to 28.35ml water	87.2	0.26	1.7	3.9	7.5	69	290
46	**Cow & Gate Premium**	Analysis and manufacturer's data (Cow & Gate)	1.6	1.76	11.2	29.5	58.0	528	2210
47	*reconstituted*	Calcd. as 1 scoop powder (4.2g) to 30ml water	87.9	0.22	1.4	3.6	7.1	64	271
48	**Farley's Oster Milk**	Analysis and manufacturer's data (Farley's)	3.1	1.81	11.5	30.2	57.8	535	2238
49	*reconstituted*	Calcd. as 1 scoop powder (4.09g) to 28.35ml water	87.8	0.23	1.5	3.8	7.3	67	282
50	**Gold Cap SMA**	Manufacturer's data (Wyeth)	1.9	1.90	12.2	28.1	56.2	512	2146
51	*reconstituted*	Manufacturer's data (Wyeth)	87.6	0.24	1.5	3.5	7.0	64	267
	Non-whey-based modified milks								
52	**Cow & Gate Plus**	Analysis and manufacturer's data (Cow & Gate)	2.1	2.27	14.5	26.3	56.2	505	2119
53	*reconstituted*	Calcd. as 1 scoop powder (4.3g) to 30ml water	87.7	0.28	1.8	3.3	7.1	63	265
54	**Farley's Oster Milk Two**	Analysis and manufacturer's data (Farley's)	4.4	1.91	12.2	19.0	65.9[a]	470	1964
55	*reconstituted*	Calcd. as 1 scoop powder (4.4g) to 28.35ml water	87.2	0.26	1.6	2.5	8.9[a]	63	263
56	**Milumil**	Analysis and manufacturer's data (Milupa)	3.2	2.15	13.7	22.8	61.2[a]	489	2056
57	*reconstituted*	Calcd. as 1 scoop granules (4.66g) to 28.35ml water	86.3	0.30	1.9	3.2	8.6[a]	69	290
58	**White Cap SMA**	Manufacturer's data (Wyeth)	1.7	1.90	12.2	28.1	55.8	511	2140
59	*reconstituted*	Manufacturer's data (Wyeth)	87.5	0.24	1.5	3.5	6.9	63	265

[a] Including oligosaccharides from the glucose syrup/maltodextrins in the product

Infant formulas

Carbohydrate fractions and fatty acids, g per 100g
Cholesterol, mg per 100g

No. 12-	Food	Starch	Total sugars	Individual sugars						Fatty acids			Cholest-erol
				Gluc	Fruct	Galact	Sucr	Malt	Lact	Satd	Mono unsatd	Poly unsatd	
Whey-based modified milks													
44	**Aptamil**	0	56.4	0	0	0	0	0	56.4	14.5	10.4	3.1	N
45	*reconstituted*	0	7.5	0	0	0	0	0	7.5	1.9	1.4	0.4	N
46	**Cow & Gate Premium**	0	58.0	0	0	0	0	0	58.0	11.6	10.5	4.7	55
47	*reconstituted*	0	7.1	0	0	0	0	0	7.1	1.4	1.3	0.6	6
48	**Farley's Oster Milk**	0	57.8	0	0	0	0	0	57.8	12.0	13.4	3.5	N
49	*reconstituted*	0	7.3	0	0	0	0	0	7.3	1.5	1.7	0.4	N
50	**Gold Cap SMA**	0	56.2	0	0	0	0	0	56.2	12.7	9.7	4.6	31
51	*reconstituted*	0	7.0	0	0	0	0	0	7.0	1.6	1.2	0.6	4
Non-whey-based modified milks													
52	**Cow & Gate Plus**	0	56.2	0	0	0	0	0	56.2	10.4	9.8	4.3	N
53	*reconstituted*	0	7.1	0	0	0	0	0	7.1	1.3	1.2	0.5	N
54	**Farley's Oster Milk Two**	Tr	24.8[a]	0.7	0	0	0	3.7	20.4	7.1	8.7	2.3	N
55	*reconstituted*	Tr	3.3[a]	0.1	0	0	0	0.5	2.7	0.9	1.2	0.3	N
56	**Milumil**	8.0	41.0[a]	0.6	0	0	0	Tr	40.4	11.6	7.1	3.1	N
57	*reconstituted*	1.1	5.8[a]	0.1	0	0	0	Tr	5.7	1.6	1.0	0.4	N
58	**White Cap SMA**	0	55.8	0	0	0	0	0	55.8	12.7	9.7	4.6	21
59	*reconstituted*	0	6.9	0	0	0	0	0	6.9	1.6	1.2	0.6	3

[a] Not including oligosaccharides from the glucose syrup/maltodextrins in the product

29

Inorganic constituents per 100g

No. 12-	Food	Na	K	Ca	Mg	P	Fe	Cu	Zn	S	Cl	Mn	Se	I
						mg							µg	
	Whey-based modified milks													
44	**Aptamil**	150	740	450	31	250	6.3	0.35	2.8	N	250	0.1	N	31
45	*reconstituted*	19	98	59	4	33	0.8	0.05	0.4	N	33	Tr	N	4
46	**Cow & Gate Premium**	150	440	420	40	230	5.5	0.30	3.0	N	290	0.1	Tr	55
47	*reconstituted*	18	54	51	4	28	0.7	Tr	0.4	N	35	Tr	Tr	7
48	**Farley's Oster Milk**	160	500	240	38	210	5.4	0.43	3.6	N	350	Tr	(5)	35
49	*reconstituted*	20	63	30	4	26	0.7	0.05	0.4	N	44	Tr	(1)	4
50	**Gold Cap SMA**	130	510	410	53	260	5.8	0.48	4.6	N	350	0.2	N	78
51	*reconstituted*	16	63	51	7	32	0.7	0.06	0.6	N	43	Tr	N	10
	Non-whey-based modified milks													
52	**Cow & Gate Plus**	250	700	630	50	390	5.8	0.30	2.9	N	440	0.1	Tr	50
53	*reconstituted*	31	87	78	6	48	0.7	Tr	0.4	N	55	Tr	Tr	7
54	**Farley's Oster Milk Two**	180	620	440	44	350	5.3	0.4	3.7	N	360	Tr	(9)	75
55	*reconstituted*	24	83	59	5	47	0.7	0.05	0.5	N	48	Tr	(1)	10
56	**Milumil**	170	630	500	46	380	3.1	0.19	2.4	N	310	Tr	N	15
57	*reconstituted*	23	89	70	6	53	0.4	Tr	0.3	N	43	Tr	N	2.0
58	**White Cap SMA**	170	600	490	42	360	5.7	0.48	4.7	N	420	0.2	N	105
59	*reconstituted*	21	74	60	5	45	0.7	0.06	0.6	N	52	Tr	N	13

No. 12-	Food	Retinol µg	Carotene µg	Vitamin D µg	Vitamin E mg	Thiamin mg	Ribo-flavin mg	Niacin mg	Trypt 60 mg	Vitamin B6 mg	Vitamin B12 µg	Folate µg	Panto-thenate mg	Biotin µg	Vitamin C mg
Whey-based modified milks															
44	**Aptamil**	865	50	8.85	5.45	0.68	1.65	4.4	3.6	0.49	2.7	119	5.70	16.6	97
45	*reconstituted*	115	6	1.17	0.72	0.09	0.22	0.6	0.5	0.06	0.4	15	0.76	2.2	13
46	**Cow & Gate Premium**	870	77	8.70	9.48	0.30	0.80	3.0	2.6	0.30	1.6	80	2.00	12.0	60
47	*reconstituted*	105	9	1.07	1.16	0.04	0.10	0.4	0.3	0.04	0.2	9	0.25	1.5	7
48	**Farley's Oster Milk**	1000	50	13.55	4.12	0.60	1.62	5.5	3.7	0.45	3.2	43	3.50	16.5	80
49	*reconstituted*	125	6	1.71	0.52	0.08	0.20	0.7	0.5	0.06	0.4	5	0.44	2.1	10
50	**Gold Cap SMA**	865	94	10.10	9.30	1.07	1.91	4.5	3.4	0.56	1.0	57	2.50	17.3	85
51	*reconstituted*	105	12	1.25	1.15	0.13	0.24	0.6	0.4	0.07	0.1	7	0.31	2.1	11
Non-whey-based modified milks															
52	**Cow & Gate Plus**	730	77	8.40	8.58	0.30	0.80	3.0	3.4	0.30	1.5	80	2.00	12.0	60
53	*reconstituted*	91	9	1.05	1.08	0.04	0.10	0.4	0.4	0.04	0.2	10	0.25	1.5	7
54	**Farley's Oster Milk Two**	775	37	11.00	2.82	0.57	0.97	5.2	2.8	0.50	1.5	41	2.80	15.6	68
55	*reconstituted*	105	4	1.48	0.38	0.08	0.13	0.7	0.4	0.07	0.2	5	0.38	2.1	9
56	**Milumil**	765	25	8.55	5.42	0.23	1.69	1.7	3.1	0.30	1.5	36	1.70	17.5	54
57	*reconstituted*	110	3	1.21	0.77	0.03	0.24	0.3	0.4	0.04	0.2	5	0.25	2.5	8
58	**White Cap SMA**	855	93	10.00	9.00	1.04	1.48	4.4	4.1	0.50	1.0	58	2.52	18.0	85
59	*reconstituted*	105	11	1.24	1.12	0.13	0.18	0.5	0.5	0.06	0.1	7	0.31	2.2	11

Infant formulas *continued*

12-060 to 12-071

Composition of food per 100g

No. 12-	Food	Description and main data sources	Water g	Total nitrogen g	Protein g	Fat g	Carbo-hydrate g	Energy value kcal	kJ
	Soya-based modified milks								
60	**Farley's Oster Soy**	Manufacturer's data (Farley's)	3.0	2.29	14.3	27.8	53.8[a]	509	2133
61	*reconstituted*	Manufacturer's data (Farley's)	87.2	0.30	1.9	3.7	7.1[a]	67	282
62	**Formula S Soya Food**	Manufacturer's data (Cow & Gate)	2.2	2.27	14.2	28.3	56.1[a]	522	2186
63	*reconstituted*	Manufacturer's data (Cow & Gate)	87.7	0.28	1.7	3.5	6.9[a]	64	269
64	**Prosobee**	Manufacturer's data (Mead Johnson Nutritionals)	2.5	2.50	15.6	27.9	55.6[a]	516	2187
65	*reconstituted*	Manufacturer's data (Mead Johnson Nutritionals)	87.8	0.31	1.9	3.5	7.0[a]	65	274
66	**Wysoy**	Manufacturer's data (Wyeth)	1.9	2.56	16.0	27.0	52.5[a]	504	2111
67	*reconstituted*	Manufacturer's data (Wyeth)	87.1	0.33	2.1	3.5	6.8[a]	65	274
	Follow-on formulas								
68	**Farley's Junior Milk**	Manufacturer's data (Farley's)	4.0	2.22	14.2	21.4	60.0[a]	474	1993
69	*reconstituted*	Manufacturer's data (Farley's)	86.9	0.30	1.9	2.9	8.3[a]	65	273
70	**Progress**	Manufacturer's data (Wyeth)	2.5	3.13	20.0	17.7	54.4[a]	443	1865
71	*reconstituted*	Manufacturer's data (Wyeth)	85.8	0.44	2.8	2.5	7.6[a]	62	262

[a] Including oligosaccharides from the glucose syrup/maltodextrins in the product

Infant formulas *continued*

Carbohydrate fractions and fatty acids, g per 100g
Cholesterol, mg per 100g

No. 12-	Food	Starch	Total sugars	Individual sugars						Fatty acids			Cholesterol
				Gluc	Fruct	Galact	Sucr	Malt	Lact	Satd	Mono unsatd	Poly unsatd	
Soya-based modified milks													
60	**Farley's Oster Soy**	Tr	N	1.1	0	0	0	N	0	11.0	12.4	3.2	0
61	*reconstituted*	Tr	N	0.1	0	0	0	N	0	1.5	1.6	0.4	0
62	**Formula S Soya Food**	Tr	10.7[a]	3.4	0	0	0	7.3	0	11.0	10.8	5.2	0
63	*reconstituted*	Tr	1.3[a]	0.4	0	0	0	0.9	0	1.4	1.3	0.6	0
64	**Prosobee**	Tr	8.0[a]	3.9	0	0	0	4.1	0	13.8	4.8	7.8	0
65	*reconstituted*	Tr	1.0[a]	0.5	0	0	0	0.5	0	1.7	0.6	1.0	0
66	**Wysoy**	Tr	N	N	0	0	13.0	N	0	12.2	9.3	4.4	10
67	*reconstituted*	Tr	N	N	0	0	1.7	N	0	1.6	1.2	0.6	1
Follow-on formulas													
68	**Farley's Junior Milk**	Tr	37.9[a]	2.2	0	0	0	2.1	33.6	8.4	9.6	2.4	N
69	*reconstituted*	Tr	5.2[a]	0.3	0	0	0	0.3	4.6	1.1	1.3	0.3	N
70	**Progress**	Tr	N	N	0	0	0	N	40.8	8.0	5.7	3.3	30
71	*reconstituted*	Tr	N	N	0	0	0	N	5.7	1.1	0.8	0.5	4

[a] Not including oligosaccharides from the glucose syrup/maltodextrins in the product

Infant formulas *continued*

No. 12-	Food	Na	K	Ca	Mg	P	mg Fe	Cu	Zn	S	Cl	Mn	µg Se	I
	Soya-based modified milks													
60	**Farley's Oster Soy**	180	550	410	42	270	4.8	0.31	3.3	N	370	0.3	N	61
61	*reconstituted*	24	73	54	5	36	0.6	0.04	0.4	N	49	Tr	N	8
62	**Formula S Soya Food**	140	530	420	40	210	4.0	0.31	3.0	N	310	0.3	Tr	110
63	*reconstituted*	17	63	52	5	26	0.5	0.04	0.4	N	39	Tr	N	13
64	**Prosobee**	190	470	430	47	310	9.3	0.31	3.9	N	310	0.1	13	35
65	*reconstituted*	24	58	53	6	39	1.2	0.04	0.5	N	39	Tr	Tr	4
66	**Wysoy**	180	650	520	59	370	5.6	0.55	4.3	N	320	0.3	N	71
67	*reconstituted*	23	84	67	8	48	0.7	0.07	0.6	N	42	Tr	N	9
	Follow-on formulas													
68	**Farley's Junior Milk**	210	720	510	50	420	5.0	0.29	2.9	N	460	Tr	N	88
69	*reconstituted*	29	97	70	7	57	0.7	0.04	0.4	N	63	Tr	N	12
70	**Progress**	270	890	810	81	820	8.8	0.50	3.8	N	540	0.3	N	128
71	*reconstituted*	37	120	120	11	110	1.2	0.07	0.5	N	75	Tr	N	18

Infant formulas *continued*

No. 12-	Food	Retinol µg	Carotene µg	Vitamin D µg	Vitamin E mg	Thiamin mg	Ribo-flavin mg	Niacin mg	Trypt 60 mg	Vitamin B6 mg	Vitamin B12 µg	Folate µg	Panto-thenate mg	Biotin µg	Vitamin C mg
Soya-based modified milks															
60	**Farley's Oster Soy**	730	Tr	8.00	3.70	0.35	0.41	5.1	3.3	0.26	1.1	26	1.80	8.0	51
61	*reconstituted*	97	Tr	1.07	0.50	0.04	0.05	0.7	0.4	0.03	0.1	3	0.23	1.1	7
62	**Formula S Soya Food**	600	0	8.70	10.00	0.30	0.80	3.0	2.3	0.30	1.6	80	2.00	12.0	60
63	*reconstituted*	78	0	1.07	1.30	0.04	0.10	0.4	0.3	0.04	0.2	10	0.30	1.5	8
64	**Prosobee**	390	0	8.11	11.60	0.39	0.46	6.2	3.0	0.31	1.5	78	2.30	39.0	42
65	*reconstituted*	49	0	1.02	1.46	0.05	0.06	0.8	0.4	0.04	0.2	10	0.29	4.9	5
66	**Wysoy**	810	90	9.75	8.80	0.88	1.07	5.0	3.6	0.68	2.1	64	2.55	42.0	80
67	*reconstituted*	105	12	1.30	1.14	0.11	0.14	0.6	0.5	0.09	0.3	8	0.33	5.4	10
Follow-on formulas															
68	**Farley's Junior Milk**	570	(34)	7.80	3.40	0.33	1.10	4.6	3.3	0.29	1.4	50	2.60	21.0	46
69	*reconstituted*	78	(5)	1.07	0.47	0.04	0.15	0.6	0.4	0.04	0.2	7	0.35	2.9	6
70	**Progress**	860	87	9.30	9.40	1.23	1.80	4.7	3.3	0.65	1.0	80	2.49	18.7	98
71	*reconstituted*	120	12	1.30	1.31	0.17	0.25	0.7	0.5	0.09	0.1	11	0.35	2.6	14

Milk-based drinks

12-072 to 12-085

Composition of food per 100g

No. 12-	Food	Description and main data sources	Water g	Total nitrogen g	Protein g	Fat g	Carbo-hydrate g	Energy value kcal	kJ
72	**Bournvita powder**	6 samples	1.5	1.39	8.7	5.1	79.0	377	1601
73	made up with whole milk	Recipe	84.6	0.53	3.4	3.9	7.6	77	324
74	made up with semi-skimmed milk	Recipe	86.5	0.55	3.5	1.7	7.8	58	247
75	made up with skimmed milk	Recipe	87.8	0.55	3.5	0.3	7.8	45	194
76	**Build-up powder**	Manufacturer's data (Carnation/Nestlé)	3.0	3.84	24.5	0.9	65.0[a]	350	1490
77	made up with whole milk	Recipe	78.1	0.88	5.6	3.6	11.7	98	414
78	made up with semi-skimmed milk	Recipe	79.8	0.90	5.7	1.5	11.9	80	343
79	made up with skimmed milk	Recipe	81.0	0.90	5.7	0.2	11.9	69	294
80	**Cambridge Diet powder**	Manufacturer's data (Cambridge Nutrition)	(3.5)	5.74	36.6	7.0	38.4	353	1496
81	made up with water	Manufacturer's data (Cambridge Nutrition)	(87.0)	0.92	5.9	1.1	6.1	56	239
82	**Cocoa powder**	10 samples, 2 brands	3.4	3.70[b]	18.5[c]	21.7	11.5	312	1301
83	made up with whole milk	Recipe	84.6	0.55	3.4	4.2	6.8	76	320
84	made up with semi-skimmed milk	Recipe	86.5	0.57	3.5	1.9	7.0	57	243
85	made up with skimmed milk	Recipe	87.8	0.57	3.5	0.5	7.0	44	190

[a] Including oligosaccharides from the glucose syrup/maltodextrins in the product

[b] Includes 0.74g purine nitrogen

[c] (Total N − purine N) x 6.25

Milk-based drinks

Carbohydrate fractions and fatty acids, g per 100g
Cholesterol, mg per 100g

No. 12-	Food	Starch	Total sugars	Individual sugars						Fatty acids			Cholest-erol
				Gluc	Fruct	Galact	Sucr	Malt	Lact	Satd	Mono unsatd	Poly unsatd	
72	**Bournvita powder**	27.0	52.0	N	N	0	N	N	N	N	N	N	N
73	*made up with whole milk*	1.0	6.6	N	N	0	N	N	N	N	N	N	N
74	*made up with*												
75	*semi-skimmed milk*	1.0	6.8	N	N	0	N	N	N	N	N	N	N
	made up with skimmed milk	1.0	6.8	N	N	0	N	N	N	N	N	N	N
76	**Build-up powder**	Tr	54.7[ab]	1.1	0	0	17.9	1.0	34.7	0.6	0.3	Tr	12
77	*made up with whole milk*	Tr	10.5	0.1	0	0	2.1	0.1	8.2	2.2	1.0	0.1	13
78	*made up with*												
	semi-skimmed milk	Tr	10.7	0.1	0	0	2.1	0.1	8.4	0.9	0.4	Tr	7
79	*made up with skimmed milk*	Tr	10.7	0.1	0	0	2.1	0.1	8.4	0.1	0.1	Tr	3
80	**Cambridge Diet powder**	0	38.4	0	0	0	0	0	38.4	N	N	N	Tr
81	*made up with water*	0	6.1	0	0	0	0	0	6.1	N	N	N	Tr
82	**Cocoa powder**	11.5	Tr	0	0	0	0	0	0	12.8	7.2	0.6	0
83	*made up with whole milk*	0.2	6.6	0	0	0	2.0	0	4.6	2.6	1.2	0.1	13
84	*made up with*												
	semi-skimmed milk	0.2	6.8	0	0	0	2.0	0	4.8	1.2	0.6	0.1	6
85	*made up with skimmed milk*	0.2	6.8	0	0	0	2.0	0	4.8	0.3	0.2	Tr	1

a Not including oligosaccharides from the glucose syrup/maltodextrins in the product
b The individual sugars differ according to flavour

Milk-based drinks

12-072 to 12-085

Inorganic constituents per 100g

No. 12-	Food	mg											µg	
		Na	K	Ca	Mg	P	Fe	Cu	Zn	S	Cl	Mn	Se	I
72	**Bournvita powder**	460	380	93	110	350	1.9	0.50	1.1	N	N	N	N	N
73	*made up with whole milk*	70	150	110	14	100	0.1	0.02	0.4	N	N	N	N	N
74	*made up with*													
	semi-skimmed milk	70	160	120	14	100	0.1	0.02	0.4	N	N	N	N	N
75	*made up with skimmed milk*	70	160	120	15	100	0.1	0.02	0.4	N	N	N	N	N
76	**Build-up powder**	380	1100	800	190	650	11.0	4.90	7.6	N	700	2.6	N	150
77	*made up with whole milk*	92	250	190	32	160	1.3	0.56	1.1	N	170	0.3	N	31
78	*made up with*													
	semi-skimmed milk	92	260	200	32	160	1.3	0.56	1.2	N	170	0.3	N	31
79	*made up with skimmed milk*	92	260	200	32	160	1.3	0.56	1.2	N	170	0.3	N	31
80	**Cambridge Diet powder**	1300	1700	700	350	700	15.7	2.60	13.1	N	1200	3.5	109	131
81	*made up with water*	210	290	110	56	110	2.5	0.40	2.1	N	210	0.5	17	21
82	**Cocoa powder**	950	1500	130	520	660	10.5	3.90	6.9	N	460	N	N	N
83	*made up with whole milk*	70	160	110	20	100	0.2	0.07	0.5	N	100	N	N	N
84	*made up with*													
	semi-skimmed milk	70	170	120	20	100	0.2	0.07	0.5	N	100	N	N	N
85	*made up with skimmed milk*	70	170	120	21	100	0.2	0.07	0.5	N	100	N	N	N

No. 12-	Food	Retinol μg	Carotene μg	Vitamin D μg	Vitamin E mg	Thiamin mg	Ribo- flavin mg	Niacin mg	Trypt 60 mg	Vitamin B6 mg	Vitamin B12 μg	Folate μg	Panto- thenate mg	Biotin μg	Vitamin C mg
72	**Bournvita powder**	Tr	Tr	Tr	Tr	N	N	N	2.0	N	Tr	N	N	N	0
73	*made up with whole milk*	50	20	0.03	0.07	N	N	N	0.8	N	0.4	N	N	N	Tr
74	*made up with semi-skimmed milk*	20	8	0.01	0.02	N	N	N	0.8	N	0.4	N	N	N	1
75	*made up with skimmed milk*	Tr	Tr	Tr	Tr	N	N	N	0.8	N	0.4	N	N	N	1
76	**Build-up powder**	660	Tr	10.00	10.00	1.50	1.50	20.0	5.8	3.00	3.0	400	13.00	55.0	60
77	*made up with whole milk*	120	18	1.17	1.23	0.21	0.32	2.4	1.3	0.40	0.7	51	1.80	8.0	8
78	*made up with semi-skimmed milk*	94	7	1.16	1.17	0.21	0.33	2.4	1.3	0.40	0.7	51	1.78	8.1	8
79	*made up with skimmed milk*	76	Tr	1.15	1.15	0.21	0.33	2.4	1.3	0.40	0.7	51	1.78	8.1	8
80	**Cambridge Diet powder**	N[a]	N	9.00	8.70	1.30	1.50	16.6	8.6	1.90	3.0	349	6.10	174.0	52
81	*made up with water*	N[b]	N	1.25	1.38	0.21	0.25	2.6	1.4	0.29	0.4	56	0.96	28.0	8
82	**Cocoa powder**	0	(40)	0	0.68	0.16	0.06	1.7	3.9	0.07	0	38	N	N	0
83	*made up with whole milk*	50	(20)	0.03	0.08	0.04	0.15	0.1	0.8	0.05	0.4	5	0.30	1.8	Tr
84	*made up with semi-skimmed milk*	20	(9)	0.01	0.04	0.04	0.16	0.1	0.8	0.05	0.4	5	0.28	1.9	1
85	*made up with skimmed milk*	Tr	Tr	Tr	0.01	0.04	0.16	0.1	0.8	0.05	0.4	5	0.28	1.9	1

a Contains 900μg retinol equivalents per 100g

b Contains 130μg retinol equivalents per 100g

Milk-based milks *continued*

Composition of food per 100g

No. 12-	Food	Description and main data sources	Water g	Total nitrogen g	Protein g	Fat g	Carbo-hydrate g	Energy value kcal	kJ
86	**Complan powder**, savoury	Chicken flavour, manufacturer's data (Farley's)	3.8	3.45	22.0	16.0	55.0ᵃ	438	1846
87	made up with water	Recipe	78.7	0.77	4.9	3.5	12.2	97	409
88	sweet	3 flavours, manufacturer's data (Farley's)	3.5	3.13	20.0	14.0	59.7ᵃ	430	1813
89	made up with water	Recipe	78.3	0.70	4.5	3.1	13.4	96	407
90	made up with whole milk	Recipe	69.3	1.08	6.9	6.1	16.9	145	612
91	made up with semi-skimmed milk	Recipe	70.8	1.09	7.0	4.3	17.0	130	550
92	made up with skimmed milk	Recipe	71.9	1.09	7.0	3.1	17.0	120	507
93	**Drinking chocolate powder**	10 tins, 3 brands	2.1	1.04ᵇ	5.5ᶜ	6.0	77.4	366	1554
94	made up with whole milk	Recipe	80.9	0.54	3.4	4.1	10.6	90	377
95	made up with semi-skimmed milk	Recipe	82.8	0.56	3.5	1.9	10.8	71	304
96	made up with skimmed milk	Recipe	83.9	0.56	3.5	0.6	10.8	59	253
97	**Horlicks powder**	Manufacturer's data (Beechams)	2.5	1.98	12.4	4.0	78.0	378	1607
98	made up with whole milk	Recipe	78.6	0.66	4.2	3.9	12.7	99	419
99	made up with semi-skimmed milk	Recipe	80.4	0.68	4.3	1.9	12.9	81	347
100	made up with skimmed milk	Recipe	81.5	0.68	4.3	0.5	12.9	70	298

ᵃ Including oligosaccharides from the glucose syrup/maltodextrins in the product

ᵇ Includes 0.16g purine nitrogen

ᶜ (Total N − purine N) x 6.25

Milk-based drinks *continued*

No. 12-	Food	Starch	Total sugars	Individual sugars						Fatty acids			Cholest-erol
				Gluc	Fruct	Galact	Sucr	Malt	Lact	Satd	Mono unsatd	Poly unsatd	
86	**Complan powder**, savoury	8.8	13.7[a]	3.3	0	0	0	3.0	7.3	7.5	6.0	1.8	N
87	*made up with water*	2.0	3.0	0.7	0	0	0	0.7	1.6	1.7	1.3	0.4	N
88	sweet	Tr	46.6[ab]	7.0	0	0	9.1	1.1	29.4	5.9[c]	6.0[c]	1.5[c]	N
89	*made up with water*	Tr	10.5	1.6	0	0	2.0	0.2	6.6	1.3	1.3	0.3	N
90	*made up with whole milk*	Tr	14.0	1.5	0	0	2.0	0.2	10.2	3.2	2.2	0.4	N
91	*made up with semi-skimmed milk*	Tr	14.1	1.5	0	0	2.0	0.2	10.4	2.1	1.7	0.4	N
92	*made up with skimmed milk*	Tr	14.1	1.5	0	0	2.0	0.2	10.4	1.3	1.3	0.3	N
93	**Drinking chocolate powder**	3.6	73.8	0	0	0	73.8	0	0	3.5	2.0	0.2	0
94	*made up with whole milk*	0.3	10.3	0	0	0	5.9	0	4.4	2.5	1.2	0.1	12
95	*made up with semi-skimmed milk*	0.3	10.5	0	0	0	5.9	0	4.6	1.2	0.6	0.1	6
96	*made up with skimmed milk*	0.3	10.5	0	0	0	5.9	0	4.6	0.3	0.2	Tr	1
97	**Horlicks powder**	25.0	53.0	N	N	N	N	N	N	N	N	N	N
98	*made up with whole milk*	2.7	10.0	N	N	0	N	N	N	N	N	N	N
99	*made up with semi-skimmed milk*	2.7	10.2	N	N	0	N	N	N	N	N	N	N
100	*made up with skimmed milk*	2.7	10.2	N	N	0	N	N	N	N	N	N	N

[a] Not including oligosaccharides from the glucose syrup/maltodextrins in the product

[b] The individual sugars differ according to flavour

[c] Chocolate flavour has a slightly different fatty acid composition

Milk-based drinks *continued*

Inorganic constituents per 100g

No. 12-	Food	Na	K	Ca	Mg	P	Fe (mg)	Cu	Zn	S	Cl	Mn	Se (μg)	I
86	**Complan powder**, savoury	1800	650	310	5	360	6.5	0.54	6.5	N	1700	0.9	N	61
87	made up with water	400	140	68	1	79	1.4	0.12	1.4	N	380	0.2	N	13
88	sweet	290	950	710	76	590	6.5	0.54	6.5	N	640	0.9	N	61
89	made up with water	65	210	160	17	130	1.5	0.12	1.5	N	140	0.2	N	14
90	made up with whole milk	110	320	250	25	200	1.5	0.12	1.7	N	220	0.2	N	25
91	made up with semi-skimmed milk	110	330	250	25	200	1.5	0.12	1.7	N	220	0.2	N	25
92	made up with skimmed milk	110	330	250	26	200	1.5	0.12	1.7	N	220	0.2	N	25
93	**Drinking chocolate powder**	250	410	33	150	190	2.4	1.10	1.9	N	130	N	N	N
94	made up with whole milk	70	160	110	22	99	0.2	0.09	0.5	N	100	N	N	N
95	made up with semi-skimmed milk	70	170	110	22	100	0.2	0.09	0.5	N	100	N	N	N
96	made up with skimmed milk	70	170	110	23	100	0.2	0.09	0.5	N	100	N	N	N
97	**Horlicks powder**	460	670	430	50	300	1.4	0.30	1.3	N	N	N	N	N
98	made up with whole milk	98	200	150	15	110	0.2	0.03	0.5	N	N	N	N	N
99	made up with semi-skimmed milk	98	210	150	15	120	0.2	0.03	0.5	N	N	N	N	N
100	made up with skimmed milk	98	210	150	16	120	0.2	0.03	0.5	N	N	N	N	N

Milk-based drinks *continued*

No. 12-	Food	Retinol µg	Carotene µg	Vitamin D µg	Vitamin E mg	Thiamin mg	Ribo-flavin mg	Niacin mg	Trypt 60 mg	Vitamin B6 mg	Vitamin B12 µg	Folate µg	Panto-thenate mg	Biotin µg	Vitamin C mg
86	**Complan powder**, savoury	430	Tr	2.20	5.20	0.77	0.87	8.7	5.2	0.95	2.2	150	3.00	33.0	21
87	*made up with water*	95	Tr	0.49	1.15	0.17	0.18	1.9	1.1	0.21	0.5	33	0.67	7.3	5
88	sweet	430	Tr	2.20	5.20	0.77	0.87	8.7	4.7	0.95	2.2	170	3.00	33.0	43
89	*made up with water*	96	Tr	0.49	1.17	0.18	0.20	1.9	1.1	0.21	0.5	38	0.67	7.4	10
90	*made up with whole milk*	135	16	0.51	1.21	0.20	0.32	2.0	1.6	0.26	0.8	42	0.93	8.7	10
91	*made up with semi-skimmed milk*	110	7	0.49	1.17	0.20	0.33	2.0	1.6	0.26	0.8	42	0.91	8.8	11
92	*made up with skimmed milk*	95	Tr	0.48	1.14	0.20	0.33	2.0	1.6	0.26	0.8	42	0.91	8.8	11
93	**Drinking chocolate powder**	0	N	0	0.18	0.06	0.04	0.5	1.2	0.02	0	10	N	N	0
94	*made up with whole milk*	47	N	0.03	0.08	0.04	0.14	0.1	0.8	0.05	0.3	5	N	N	Tr
95	*made up with semi-skimmed milk*	19	N	0.01	0.04	0.04	0.15	0.1	0.8	0.05	0.3	5	N	N	1
96	*made up with skimmed milk*	Tr	N	Tr	0.02	0.04	0.15	0.1	0.8	0.05	0.3	5	N	N	1
97	**Horlicks powder**	625	0	2.08	N	1.00	1.33	15.0	3.0	N	N	N	N	N	0
98	*made up with whole milk*	115	18	0.25	N	0.14	0.28	1.7	1.0	N	N	N	N	N	Tr
99	*made up with semi-skimmed milk*	86	8	0.23	N	0.14	0.29	1.7	1.0	N	N	N	N	N	1
100	*made up with skimmed milk*	68	Tr	0.23	N	0.14	0.29	1.7	1.0	N	N	N	N	N	1

Milk-based drinks *continued*

Composition of food per 100g

No. 12-	Food	Description and main data sources	Water g	Total nitrogen g	Protein g	Fat g	Carbo-hydrate g	Energy value kcal	kJ
101	**Microdiet powder**	For 'micro-drink', manufacturer's data (Uni-Vite)	N	6.72	42.0	3.0	35.0	326	1385
102	*made up with water*	Recipe	N	0.78	4.9	0.3	4.1	38	161
103	**Milk shake,** *purchased*	21 samples, thick take-away type	80.0	0.46	2.9	3.2	13.2	90	379
104	**Milk shake powder**	6 samples (Nesquik), 3 flavours	0.5	0.21	1.3	1.6	98.3	388	1654
105	*made up with whole milk*	Recipe	81.9	0.48	3.1	3.7	11.1	87	368
106	*made up with semi-skimmed milk*	Recipe	83.7	0.50	3.2	1.6	11.3	69	294
107	*made up with skimmed milk*	Recipe	85.0	0.50	3.2	0.2	11.3	57	242
108	**Ovaltine powder**	Manufacturer's data (Wander)	2.0	1.41	9.0	2.7	79.4	358	1523
109	*made up with whole milk*	Recipe	78.5	0.60	3.8	3.8	12.9	97	410
110	*made up with semi-skimmed milk*	Recipe	80.3	0.62	3.9	1.7	13.0	79	338
111	*made up with skimmed milk*	Recipe	81.5	0.62	3.9	0.4	13.0	68	289

Milk-based drinks *continued*

Carbohydrate fractions and fatty acids, g per 100g
Cholesterol, mg per 100g

No. 12-	Food	Starch	Total sugars	Individual sugars						Fatty acids			Cholest-erol
				Gluc	Fruct	Galact	Sucr	Malt	Lact	Satd	Mono unsatd	Poly unsatd	
101	**Microdiet powder**	N	N	N	0	0	N	N	N	N	N	N	N
102	*made up with water*	N	N	N	0	0	N	N	N	N	N	N	N
103	**Milk shake**, *purchased*	Tr	13.2	1.2	1.1	0	6.2	0	4.7	(2.0)	(0.9)	(0.1)	10
104	**Milk shake powder**	Tr	98.3	0.1	0	0	95.2	2.8	0.2	N	N	N	Tr
105	*made up with whole milk*	Tr	11.1	Tr	0	0	6.5	0.2	4.5	N	N	N	13
106	*made up with semi-skimmed milk*	Tr	11.3	Tr	0	0	6.5	0.2	4.7	N	N	N	6
107	*made up with skimmed milk*	Tr	11.3	Tr	0	0	6.5	0.2	4.7	N	N	N	1
108	**Ovaltine powder**	N	N	N	N	0	N	N	N	N	N	N	N
109	*made up with whole milk*	N	N	N	N	0	N	N	N	N	N	N	N
110	*made up with semi-skimmed milk*	N	N	N	N	0	N	N	N	N	N	N	N
111	*made up with skimmed milk*	N	N	N	N	0	N	N	N	N	N	N	N

Milk-based drinks *continued*

Inorganic constituents per 100g

No. 12-	Food	Na	K	Ca	Mg	P	mg Fe	Cu	Zn	S	Cl	Mn	µg Se	I
101	**Microdiet powder**	1500	2010	900	350	800	20.0	2.00	20.0	N	1800	2.9	60	150
102	*made up with water*	170	230	100	40	93	2.3	0.23	2.3	N	210	0.3	7	17
103	**Milk shake,** *purchased*	55	130	100	9	86	1.0	0.01	0.4	N	91	Tr	N	N
104	**Milk shake powder**	20	150	8	27	53	2.0	0.10	0.4	N	27	0.2	N	32
105	*made up with whole milk*	52	140	110	12	89	0.2	0.01	0.4	N	95	Tr	N	16
106	*made up with semi-skimmed milk*	52	150	110	12	92	0.2	0.01	0.4	N	95	Tr	N	16
107	*made up with skimmed milk*	52	150	110	13	92	0.2	0.01	0.4	N	95	Tr	N	16
108	**Ovaltine powder**	160	640	83	96	430	1.9	1.00	1.2	14	210	N	N	N
109	*made up with whole milk*	66	190	110	20	130	0.3	0.11	0.5	28	110	N	N	N
110	*made up with semi-skimmed milk*	66	200	110	20	130	0.3	0.11	0.5	29	110	N	N	N
111	*made up with skimmed milk*	66	200	110	21	130	0.3	0.11	0.5	29	110	N	N	N

Milk-based drinks *continued*

No. 12-	Food	Retinol µg	Carotene µg	Vitamin D µg	Vitamin E mg	Thiamin mg	Ribo-flavin mg	Niacin mg	Trypt 60 mg	Vitamin B6 mg	Vitamin B12 µg	Folate µg	Panto-thenate mg	Biotin µg	Vitamin C mg
101	**Microdiet powder**	1000	N[a]	3.00	10.00	2.00	2.00	19.0	9.5	3.00	3.0	400	7.00	200.0	70
102	*made up with water*	115	N	0.35	1.17	0.23	0.23	2.2	1.1	0.35	0.3	46	0.82	23.3	8
103	**Milk shake**, *purchased*	21	51	Tr	0.05	0.03	0.18	0.1	0.7	0.05	0.3	4	0.31	1.9	1
104	**Milk shake powder**	Tr	Tr	0	0.15	Tr	0.02	0.2	0.3	0.01	0	3	N	N	0
105	*made up with whole milk*	48	19	0.03	0.09	0.04	0.16	0.1	0.7	0.06	0.4	5	N	N	1
106	*made up with semi-skimmed milk*	19	8	0.01	0.04	0.04	0.17	0.1	0.7	0.06	0.4	5	N	N	1
107	*made up with skimmed milk*	Tr	Tr	Tr	0.01	0.04	0.17	0.1	0.7	0.06	0.4	5	N	N	1
108	**Ovaltine powder**	625	Tr	2.10	N	1.00	1.30	15.0	2.1	N	1.7	N	N	N	0
109	*made up with whole milk*	114	18	0.25	N	0.14	0.28	1.7	0.9	N	0.5	N	N	N	Tr
110	*made up with semi-skimmed milk*	86	8	0.24	N	0.14	0.29	1.7	0.9	N	0.5	N	N	N	1
111	*made up with skimmed milk*	68	Tr	0.23	N	0.14	0.29	1.7	0.9	N	0.5	N	N	N	1

[a] β-Carotene is added as a colouring agent in some flavours

47

Creams

Composition of food per 100g

No. 12-	Food	Description and main data sources	Water g	Total nitrogen g	Protein g	Fat g	Carbo-hydrate g	Energy value kcal	kJ
Fresh creams (pasteurised)									
112	**Half**	10 samples, 5 brands	78.9	0.47	3.0	13.3	4.3	148	613
113	**Single**	10 samples, 5 brands	73.7	0.41	2.6	19.1	4.1	198	817
114	**Soured**	8 samples, 4 brands	72.5	0.45	2.9	19.9	3.8	205	845
115	**Whipping**	10 samples, 6 brands	55.4	0.31	2.0	39.3	3.1	373	1539
116	**Double**	12 samples, 5 brands	47.5	0.27	1.7	48.0	2.7	449	1849
117	**Clotted**	17 samples, 3 brands	32.2	0.25	1.6	63.5	2.3	586	2413
Frozen creams (pasteurised)									
118	**Single**	4 samples	72.9	0.48	3.1	20.0	4.0	207	856
119	**Whipping**	10 samples, 2 brands	55.1	0.30	1.9	40.0	3.0	379	1560
Sterilised creams									
120	**Sterilised**, canned	13 cans, 6 brands	69.2	0.39	2.5	23.9	3.7	239	985
UHT creams									
121	**Half**	5 samples (Sainsbury's)	79.1	0.44	2.8	12.3	4.3	138	572
122	**Single**	8 samples, 4 brands	73.9	0.41	2.6	19.0	4.0	196	811
123	**Whipping**	10 samples, 5 brands	57.9	(0.31)	(2.0)	(39.3)	(3.1)	(373)	(1539)
124	**Canned spray**	8 samples (Anchor)	58.4	0.30	1.9	32.0	3.5	309	1273

Carbohydrate fractions and fatty acids, g per 100g
Cholesterol, mg per 100g

No. 12-	Food	Starch	Total sugars	Individual sugars						Fatty acids			Cholest- erol
				Gluc	Fruct	Galact	Sucr	Malt	Lact	Satd	Mono unsatd	Poly unsatd	
Fresh creams (pasteurised)													
112	Half	0	4.3	0	0	0	0	0	4.3	8.3	3.9	0.4	40
113	Single	0	4.1	0	0	0	0	0	4.1	11.9	5.5	0.5	55
114	Soured	0	3.8	0	0	0	0	0	3.8	12.5	5.8	0.6	60
115	Whipping	0	3.1	0	0	0	0	0	3.1	24.6	11.4	1.1	105
116	Double	0	2.7	0	0	0	0	0	2.7	30.0	13.9	1.4	130
117	Clotted	0	2.3	0	0	0	0	0	2.3	39.7	18.4	1.8	170
Frozen creams (pasteurised)													
118	Single	0	4.0	0	0	0	0	0	4.0	12.5	5.8	0.6	60
119	Whipping	0	3.0	0	0	0	0	0	3.0	25.0	11.6	1.2	110
Sterilised creams													
120	Sterilised, canned	0	3.7	0	0	0	0	0	3.7	14.9	6.9	0.7	65
UHT creams													
121	Half	0	4.3	0	0	0	0	0	4.3	7.7	3.6	0.4	35
122	Single	0	4.0	0	0	0	0	0	4.0	11.9	5.5	0.5	55
123	Whipping	0	(3.1)	0	0	0	0	0	(3.1)	(24.6)	(11.4)	(1.1)	105
124	Canned spray	0	3.5	0	0	0	0	0	3.5	20.0	9.3	0.9	85

No. 12-	Food	Na	K	Ca	Mg	P	Fe	Cu	Zn	S	Cl	Mn	Se	I
							mg						µg	
													Se	I

Fresh creams (pasteurised)

No. 12-	Food	Na	K	Ca	Mg	P	Fe	Cu	Zn	S	Cl	Mn	Se	I
112	**Half**	49	120	99	11	82	0.1	Tr	0.3	28	77	Tr	Tr	N
113	**Single**	49	120	91	9	76	0.1	Tr	0.5	24	80	Tr	Tr	N
114	**Soured**	41	110	93	10	81	0.4	Tr	0.5	27	81	Tr	Tr	N
115	**Whipping**	40	80	62	7	58	Tr	Tr	0.3	22	59	Tr	Tr	N
116	**Double**	37	65	50	6	50	0.2	Tr	0.2	16	51	Tr	Tr	Tr
117	**Clotted**	18	55	37	5	40	0.1	0.09	0.2	15	40	Tr	Tr	Tr

Frozen creams (pasteurised)

| 118 | **Single** | 38 | 120 | 90 | 9 | 78 | Tr | Tr | 0.3 | 29 | 85 | Tr | Tr | N |
| 119 | **Whipping** | 33 | 89 | 60 | 7 | 56 | Tr | Tr | 0.3 | 18 | 59 | Tr | Tr | N |

Sterilised creams

| 120 | **Sterilised**, canned | 53 | 110 | 86 | 10 | 73 | 0.8 | Tr | 1.1 | 23 | 78 | Tr | Tr | N |

UHT creams

121	**Half**	46	130	96	9	78	Tr	0.01	0.4	26	86	Tr	Tr	N
122	**Single**	45	150	85	9	77	0.1	Tr	0.4	24	80	Tr	Tr	N
123	**Whipping**	33	74	63	6	60	0.4	Tr	0.2	19	73	Tr	Tr	N
124	**Canned spray**	33	92	66	7	57	1.0	Tr	0.4	18	62	Tr	Tr	N

No. 12-	Food	Retinol µg	Carotene µg	Vitamin D µg	Vitamin E mg	Thiamin mg	Ribo-flavin mg	Niacin mg	Trypt 60 mg	Vitamin B6 mg	Vitamin B12 µg	Folate µg	Panto-thenate mg	Biotin µg	Vitamin C mg
Fresh creams (pasteurised)															
112	**Half**	190	54	0.10	0.29	0.03	0.18	0.07	0.71	0.05	0.3	6	0.26	1.7	1
113	**Single**	315	125	0.14	0.40	0.04	0.17	0.07	0.61	0.05	0.3	7	0.28	1.8	1
114	**Soured**	330	105	0.15	0.44	0.03	0.17	0.07	0.67	0.04	0.2	12	0.24	1.5	Tr
115	**Whipping**	565	265	0.22	0.86	0.02	0.17	0.04	0.47	0.04	0.2	7	0.22	1.4	1
116	**Double**	600	325	0.27	1.10	0.02	0.16	0.04	0.41	0.03	0.2	7	0.19	1.1	1
117	**Clotted**	705	685	0.28	1.48	0.02	0.16	0.04	0.37	0.03	0.1	6	0.14	1.0	Tr
Frozen creams (pasteurised)															
118	**Single**	330	130	0.15	0.44	0.03	0.15	0.06	0.72	0.04	0.4	8	0.29	2.5	Tr
119	**Whipping**	575	270	0.22	0.88	0.03	0.17	0.03	0.45	0.03	0.3	8	0.19	1.5	1
Sterilised creams															
120	**Sterilised**, canned	240	215	Tr	0.48	0.02	0.16	0.06	0.59	0.02	0.1	1	0.25	2.1	Tr
UHT creams															
121	**Half**	165	120	(0.10)	0.15	0.03	0.17	0.08	0.66	0.04	0.2	3	0.28	1.8	Tr
122	**Single**	255	180	(0.14)	0.23	0.03	0.17	0.07	0.61	0.04	0.2	3	0.30	1.9	Tr
123	**Whipping**	(565)	(265)	(0.22)	0.86	0.03	0.16	0.05	0.51	0.03	0.2	1	0.23	1.4	Tr
124	**Canned spray**	370	340	(0.18)	0.74	0.03	0.17	0.05	0.45	0.03	0.2	1	0.19	1.7	0

Creams *continued*

Composition of food per 100g

No. 12-	Food	Description and main data sources	Water g	Total nitrogen g	Protein g	Fat g	Carbo-hydrate g	Energy value kcal	kJ
Imitation creams									
125	**Dessert Top**	Manufacturer's data (Carnation/Nestlé)	N	0.38	2.4	28.8	6.0	291	1202
126	**Elmlea**, single	Analysis and manufacturer's data (Van den Berghs)	N	0.47	3.0	19.0	3.5	195	810
127	whipping	Analysis and manufacturer's data (Van den Berghs)	N	0.38	2.4	33.0	2.3	315	1299
128	double	Analysis and manufacturer's data (Van den Berghs)	N	0.38	2.4	48.0	2.3	450	1853
129	**Smatana**	Creamed cultured dairy product	N	0.74	4.7	10.0	5.6	130	540
130	**Tip Top**	Manufacturer's data (Nestlé)	N	0.78	5.0	6.5	8.5	110	460

Creams continued

Carbohydrate fractions and fatty acids, g per 100g
Cholesterol, mg per 100g

Imitation creams

| No. 12- | Food | Starch | Total sugars | Individual sugars | | | | | | Fatty acids | | | Cholest-erol |
				Gluc	Fruct	Galact	Sucr	Malt	Lact	Satd	Mono unsatd	Poly unsatd	
125	**Dessert Top**	Tr	N	N	0	0	N	N	N	27.0	N	N	Tr
126	**Elmlea**, single	0	3.5	0	0	0	0	0	3.5	15.3	2.6	0.3	30
127	whipping	0	2.3	0	0	0	0	0	2.3	27.2	3.7	0.5	31
128	double	0	2.3	0	0	0	0	0	2.3	30.2	12.3	3.5	33
129	**Smatana**	0	5.6	0	0	Tr	0	0	5.6	6.3	2.9	0.3	(27)
130	**Tip Top**	N	N	0	0	0	N	0	N	5.8	N	N	Tr

Inorganic constituents per 100g

No. 12-	Food	Na	K	Ca	Mg	P	Fe	Cu	Zn	S	Cl	Mn	Se	I
							mg						µg	
Imitation creams														
125	**Dessert Top**	50	100	80	8	70	N	N	N	N	N	N	N	N
126	**Elmlea**, single	54	120	110	N	95	N	N	N	N	N	N	N	N
127	whipping	42	93	88	N	74	N	N	N	N	N	N	N	N
128	double	42	93	88	N	74	N	N	N	N	N	N	N	N
129	**Smatana**	N	N	N	N	N	N	Tr	N	N	N	Tr	Tr	N
130	**Tip Top**	90	220	160	16	150	N	N	N	N	N	N	N	N

Creams continued

Imitation creams

No. 12-	Food	Retinol µg	Carotene µg	Vitamin D µg	Vitamin E mg	Thiamin mg	Ribo-flavin mg	Niacin mg	Trypt 60 mg	Vitamin B6 mg	Vitamin B12 µg	Folate µg	Panto-thenate mg	Biotin µg	Vitamin C mg
125	**Dessert Top**	Tr	400	Tr	N	N	N	N	0.57	N	N	N	N	N	Tr
126	**Elmlea**, single	N	N	N	N	N	N	N	0.71	N	N	N	N	N	N
127	whipping	N	N	N	N	N	N	N	0.54	N	N	N	N	N	N
128	double	N	N	N	N	N	N	N	0.24	N	N	N	N	N	N
129	**Smatana**	(165)	(65)	(0.07)	(0.21)	N	N	N	1.1	N	N	N	N	N	(1)
130	**Tip Top**	Tr	6	Tr	N	N	N	N	1.17	N	N	N	N	N	Tr

Cheeses

Composition of food per 100g

No.	Food	Description and main data sources	Water g	Total nitrogen g	Protein g	Fat g	Carbo-hydrate g	Energy value kcal	Energy value kJ
12-									
131	**Brie**	10 samples	48.6	3.02	19.3	26.9	Tr	319	1323
132	**Caerphilly**	10-20 samples	41.8	3.64	23.2	31.3	0.1	375	1554
133	**Camembert**	10 samples	50.7	3.27	20.9	23.7	Tr	297	1232
134	**Cheddar**, *average*	Weighted average from 5 countries	36.0	4.00	25.5	34.4	0.1	412	1708
135	Australian	10 samples	36.1	3.99	24.9	34.3	0.1	409	1694
136	Canadian	10 samples	34.3	4.11	26.2	35.3	0.1	423	1753
137	English	10-15 samples	36.0	4.00	25.5	34.4	0.1	412	1708
138	Irish	10 samples	37.0	3.94	25.1	33.8	0.1	405	1679
139	New Zealand	10 samples	34.1	4.12	26.3	35.4	0.1	424	1759
140	vegetarian	10 samples	33.9	4.04	25.8	35.7	Tr	425	1759
141	**Cheddar-type**, *reduced fat*	10 samples, Tendale	47.1	4.94	31.5	15.0	Tr	261	1091
142	**Cheese spread**, plain	10 samples, 3 brands	53.3	2.11	13.5	22.8[a]	4.4	276	1143
143	flavoured	10 samples, assorted flavours	56.8	2.23	14.2	20.5	4.4	258	1070
144	lactic	10 samples (St Ivel)	52.2	2.78	17.7	24.5	*1.8*	*298*	*1236*
145	**Cheshire**	10-33 samples	40.6	3.76	24.0	31.4	0.1	379	1571
146	**Cheshire-type**, *reduced fat*	10 samples, Tendale	46.9	5.12	32.7	15.3	Tr	269	1122

[a] Reduced fat varieties contain approximately 9.0g fat per 100g

Cheeses

Carbohydrate fractions and fatty acids, g per 100g
Cholesterol, mg per 100g

| No. 12- | Food | Starch | Total sugars | Individual sugars | | | | | | Fatty acids | | | Cholest-erol |
				Gluc	Fruct	Galact	Sucr	Malt	Lact	Satd	Mono unsatd	Poly unsatd	
131	**Brie**	0	Tr	0	0	0	0	0	Tr	(16.8)	(7.8)	(0.8)	100
132	**Caerphilly**	0	0.1	0	0	0	0	0	0.1	19.6	9.1	0.9	90
133	**Camembert**	0	Tr	0	0	0	0	0	Tr	(14.8)	(6.9)	(0.7)	75
134	**Cheddar**, *average*	0	0.1	0	0	0	0	0	0.1	21.7	9.4	1.4	100
135	Australian	0	0.1	0	0	0	0	0	0.1	(21.6)	(9.4)	(1.4)	(100)
136	Canadian	0	0.1	0	0	0	0	0	0.1	(22.2)	(9.7)	(1.4)	(100)
137	English	0	0.1	0	0	0	0	0	0.1	21.7	9.4	1.4	100
138	Irish	0	0.1	0	0	0	0	0	0.1	(21.3)	(9.3)	(1.4)	(100)
139	New Zealand	0	0.1	0	0	0	0	0	0.1	(22.3)	(9.7)	(1.4)	(100)
140	vegetarian	0	Tr	0	0	0	0	0	Tr	22.5	9.8	1.5	105
141	**Cheddar-type**, *reduced fat*	0	Tr	0	0	0	0	0	Tr	9.4	4.4	0.4	43
142	**Cheese spread**, plain	0	4.4	0	0	0	0	0	4.4	14.3	6.6	0.7	(65)
143	flavoured	0	4.4	0	0	0	0	0	4.4	12.8	5.9	0.6	(60)
144	lactic	0	1.8	0	0	0	0	0	1.8	15.3	7.1	0.7	70
145	**Cheshire**	0	0.1	0	0	0	0	0	0.1	19.6	9.1	0.9	90
146	**Cheshire-type**, *reduced fat*	0	Tr	0	0	0	0	0	Tr	9.6	4.5	0.4	44

No. 12-	Food	Na	K	Ca	Mg	P	mg Fe	Cu	Zn	S	Cl	Mn	μg Se	I
131	**Brie**	700	100	540	27	390	0.8	Tr	2.2	N	1060	Tr	N	N
132	**Caerphilly**	480	91	550	20	400	0.7	0.14	3.3	230	750	Tr	(11)	(46)
133	**Camembert**	650	100	350	21	310	0.2	0.07	2.7	N	1120	Tr	N	N
134	**Cheddar**, *average*	670	77	720	25	490	0.3	0.03	2.3	230	1030	Tr	12	39
135	Australian	690	82	580	26	520	0.4	0.13	2.1	230	1030	Tr	(12)	N
136	Canadian	640	86	650	23	510	0.3	0.12	3.6	230	930	Tr	(12)	44
137	English	670	79	740	26	490	0.2	Tr	2.3	230	1010	Tr	12	46
138	Irish	710	71	700	23	480	0.5	0.02	1.9	230	1150	Tr	(12)	20
139	New Zealand	630	62	750	22	500	0.2	0.12	2.0	230	1010	Tr	(12)	58
140	vegetarian	670	67	690	31	490	0.2	Tr	1.9	N	990	0.1	(12)	(46)
141	**Cheddar-type**, *reduced fat*	670	110	840	39	620	0.2	0.05	2.8	(230)	1110	Tr	(15)	N
142	**Cheese spread**, plain	1060	240	420	25	790	0.2	0.12	2.1	N	820	Tr	(6)	29
143	flavoured	1130	140	430	23	740	0.2	0.11	2.3	N	840	Tr	(6)	(29)
144	lactic	1020	130	450	21	450	0.2	0.15	2.3	N	990	Tr	(5)	N
145	**Cheshire**	550	87	560	19	400	0.3	0.13	3.3	230	830	Tr	(11)	(46)
146	**Cheshire-type**, *reduced fat*	470	95	650	29	550	0.2	0.19	3.5	(230)	860	Tr	(15)	N

No. 12-	Food	Retinol µg	Carotene µg	Vitamin D µg	Vitamin E mg	Thiamin mg	Ribo-flavin mg	Niacin mg	Trypt 60 mg	Vitamin B6 mg	Vitamin B12 µg	Folate µg	Panto-thenate mg	Biotin µg	Vitamin C mg
131	**Brie**	285	210	0.20	0.84	0.04[a]	0.43	0.43	4.53	0.15[b]	1.2	58	0.35	5.6	Tr
132	**Caerphilly**	315	210	0.24	0.78	0.03	0.47	0.11	5.46	0.11	1.1	50	0.29	3.5	Tr
133	**Camembert**	230	315	(0.18)	0.65	0.05[a]	0.52	0.96	4.91	0.22[b]	1.1	102	0.36	7.6	Tr
134	**Cheddar**, *average*	325	225	0.26	0.53	0.03	0.40	0.07	6.00	0.10	1.1	33	0.36	3.0	Tr
135	Australian	265	305	(0.26)	0.47	0.02	0.44	0.02	5.99	0.13	0.8	42	0.35	2.1	Tr
136	Canadian	260	115	(0.26)	0.47	0.02	0.39	0.03	6.17	0.10	0.9	16	0.34	2.4	Tr
137	English	315	125	0.26	0.54	0.03	0.42	0.09	6.00	0.10	1.2	37	0.38	3.1	Tr
138	Irish	350	285	(0.26)	0.52	0.03	0.45	0.05	5.91	0.09	1.1	22	0.31	3.2	Tr
139	New Zealand	280	235	(0.26)	0.56	0.02	0.52	0.04	6.18	0.08	0.9	33	0.30	3.1	Tr
140	vegetarian	385	460	0.27	0.80	0.03	0.45	0.04	6.06	0.11	1.2	25	0.46	2.6	Tr
141	**Cheddar-type**, *reduced fat*	165	100	0.11	0.39	0.03	0.53	0.09	7.41	0.13	1.3	56	0.51	3.8	Tr
142	**Cheese spread**, plain	275	105	0.17	0.24	0.05	0.36	0.12	3.17	0.08	0.6	19	0.51	3.6	Tr
143	flavoured	(245)	(94)	0.16	(0.22)	0.15	0.25	0.16	3.35	0.08	0.6	19	0.31	2.4	Tr
144	lactic	(325)	(130)	0.19	(0.57)	0.03	0.29	0.16	4.17	0.08	0.5	31	0.26	3.5	0
145	**Cheshire**	350	220	0.24	0.70	0.03	0.48	0.11	5.64	0.09	0.9	40	0.31	4.0	Tr
146	**Cheshire-type**, *reduced fat*	150	80	0.12	0.31	0.05	0.56	0.18	7.68	0.10	1.4	58	0.51	5.0	Tr

[a] Rind 0.5mg
[b] Rind 0.4mg

Cheeses *continued*

Composition of food per 100g

No. 12-	Food	Description and main data sources	Water g	Total nitrogen g	Protein g	Fat g	Carbohydrate g	Energy value kcal	kJ
147	**Cottage cheese**, plain	10-19 samples	79.1	2.16	13.8	3.9	2.1	98	413
148	*with additions*	10 samples, mixed, e.g. with pineapple, Cheddar cheese	76.9	2.00	12.8	3.8	2.6	95	400
149	*reduced fat*	6 samples, different brands	80.2	2.08	13.3	1.4	3.3	78	331
150	**Cream cheese**	3 samples	45.5	0.49	3.1	47.4	Tr	439	1807
151	**Danish blue**	10 samples	45.3	3.15	20.1	29.6	Tr	347	1437
152	**Derby**	10-14 samples	38.0	3.79	24.2	33.9	0.1	402	1667
153	**Double Gloucester**	10-22 samples	37.3	3.90	24.6	34.0	0.1	405	1678
154	**Edam**	10 samples	43.8	4.08	26.0	25.4	Tr	333	1382
155	**Edam-type**, *reduced fat*	4 samples, 2 brands	48.2	5.11	32.6	10.9	Tr	229	957
156	**Emmental**	Literature sources	35.7	4.50	28.7	29.7	Tr	382	1587
157	**Feta**	18 samples, made from sheeps and goats milk	56.5	2.45	15.6	20.2	1.5	250	1037
158	**Fromage frais**, plain	12 samples, 3 brands	77.9	1.06	6.8	7.1	5.7	113	469
159	*fruit*	11 samples, 4 brands, mixed flavours	71.9	1.06	6.8	5.8	13.8	131	551
160	*very low fat*	10 samples, 4 brands, plain and fruit	83.7	1.21	7.7	0.2	6.8	58	247
161	**Full fat soft cheese**	e.g. Philadelphia-type. Manufacturer's data plus calculation	58.0	1.35	8.6	31.0	Tr	313	1293

Cheeses continued

Carbohydrate fractions and fatty acids, g per 100g
Cholesterol, mg per 100g

No. 12-	Food	Starch	Total sugars	Gluc	Fruct	Galact	Sucr	Malt	Lact	Satd	Mono unsatd	Poly unsatd	Cholest-erol
147	**Cottage cheese**, plain	0	2.1	0	0	0	0	0	2.1	2.4	1.1	0.1	13
148	*with additions*	0	2.6	0.6	0	0	0	0	2.0	2.4	1.1	0.1	13
149	*reduced fat*	0	3.3	0	0	0	0	0	3.3	0.9	0.4	Tr	5
150	**Cream cheese**	0	Tr	0	0	0	0	0	Tr	29.7	13.7	1.4	95
151	**Danish blue**	0	Tr	0	0	0	0	0	Tr	(18.5)	(8.6)	(0.9)	75
152	**Derby**	0	0.1	0	0	0	0	0	0.1	21.2	9.8	1.0	100
153	**Double Gloucester**	0	0.1	0	0	0	0	0	0.1	21.3	9.9	1.0	100
154	**Edam**	0	Tr	0	0	0	0	0	Tr	(15.9)	(7.4)	(0.7)	80
155	**Edam-type**, *reduced fat*	0	Tr	0	0	0	0	0	Tr	6.8	3.2	0.3	32
156	**Emmental**	0	Tr	0	0	0	0	0	Tr	(18.6)	(8.6)	(0.9)	90
157	**Feta**	0	1.5	0	0	0.1	0	0	1.4	(13.7)	(4.1)	(0.6)	70
158	**Fromage frais**, plain	0	5.7	0.1	0	0.1	2.2	0	3.3	4.4	2.1	0.2	25
159	*fruit*	0	13.8	1.1	1.1	Tr	8.8	0	2.8	3.6	1.7	0.2	21
160	*very low fat*	Tr	6.8	0.2	0.4	0.1	3.0	0	3.1	0.1	0.1	Tr	1
161	**Full fat soft cheese**	0	Tr	0	0	0	0	0	Tr	19.4	9.0	0.9	90

Cheeses continued

Inorganic constituents per 100g

No. 12-	Food	Na	K	Ca	Mg	P	Fe	Cu	Zn	S	Cl	Mn	Se	I
							mg						µg	
147	Cottage cheese, plain	380	89	73	9	160	0.1	0.04	0.6	N	550	Tr	(4)	N
148	with additions	360	130	110	12	160	0.1	0.05	0.5	N	590	Tr	(4)	N
149	reduced fat	(380)	(89)	(73)	(9)	(160)	(0.1)	(0.04)	(0.6)	N	(550)	Tr	(4)	N
150	Cream cheese	300	160	98	10	100	0.1	(0.04)	0.5	N	480	Tr	(1)	N
151	Danish blue	1260	89	500	27	370	0.2	0.08	2.0	N	1950	Tr	2	9
152	Derby	580	87	680	26	470	0.4	0.02	1.8	230	1090	Tr	(11)	(46)
153	Double Gloucester	590	79	660	23	460	0.4	0.03	1.8	230	900	Tr	(12)	(46)
154	Edam	1020	97	770	39	530	0.4	0.05	2.2	N	1570	Tr	N	N
155	Edam-type, reduced fat	N	N	N	N	N	N	N	N	N	N	Tr	N	N
156	Emmental	450	89	970	35	590	0.3	1.30	4.4	200	690	Tr	7	N
157	Feta	1440	95	360	20	280	0.2	0.07	0.9	N	2350	Tr	N	N
158	Fromage frais, plain	31	110	89	8	110	0.1	Tr	0.3	N	100	Tr	(2)	N
159	fruit	35	110	86	8	110	0.1	0.02	0.4	N	78	Tr	(2)	N
160	very low fat	(33)	(110)	(87)	(8)	(110)	(0.1)	(0.01)	(0.3)	N	(89)	Tr	(2)	N
161	Full fat soft cheese	(330)	(150)	(110)	(9)	(130)	(0.1)	(0.10)	(0.7)	N	(600)	Tr	(3)	N

No. 12-	Food	Retinol µg	Carotene µg	Vitamin D µg	Vitamin E mg	Thiamin mg	Ribo-flavin mg	Niacin mg	Trypt 60 mg	Vitamin B6 mg	Vitamin B12 µg	Folate µg	Panto-thenate mg	Biotin µg	Vitamin C mg
147	**Cottage cheese**, plain	44	10	0.03	0.08	0.03	0.26	0.13	3.24	0.08	0.7	27	0.40	3.0	Tr
148	*with additions*	43	10	0.03	0.08	0.06	0.21	0.19	3.00	0.08	0.6	13	0.31	3.0	1
149	*reduced fat*	16	4	0.01	0.03	(0.03)	(0.26)	(0.13)	3.12	(0.08)	(0.7)	(27)	(0.40)	(3.0)	Tr
150	**Cream cheese**	385	220	0.27	1.00	0.03	0.13	0.06	0.73	0.04	0.3	11	0.27	1.6	Tr
151	**Danish blue**	280	250	(0.23)	0.76	0.03	0.41	0.48	4.73	0.12	1.0	50	0.53	2.7	Tr
152	**Derby**	340	220	0.26	0.46	0.03	0.41	0.03	5.69	0.10	1.4	26	0.29	3.0	Tr
153	**Double Gloucester**	345	195	0.26	0.64	0.03	0.45	0.07	5.85	0.11	1.3	30	0.32	3.1	Tr
154	**Edam**	175	150	(0.19)	0.48	0.03	0.35	0.07	6.12	0.09	2.1	40	0.38	1.8	Tr
155	**Edam-type**, *reduced fat*	75	64	0.08	0.21	N	N	N	7.67	N	N	N	N	N	Tr
156	**Emmental**	320	140	N	0.44	0.05	0.35	0.10	6.75	0.09	2.0	20	0.40	3.0	Tr
157	**Feta**	220	33	0.50	0.37	0.04	0.21	0.19	3.45	0.07	1.1	23	0.36	2.4	Tr
158	**Fromage frais**, plain	100	Tr	0.05	0.02	0.04	0.40	0.13	1.59	0.10	1.4	15	N	N	Tr
159	*fruit*	82	N	0.04	(0.01)	0.02	0.35	0.15	1.59	0.04	1.4	15	N	N	Tr
160	*very low fat*	3	N	Tr	Tr	(0.03)	(0.37)	(0.14)	1.81	(0.07)	(1.4)	(15)	N	N	Tr
161	**Full fat soft cheese**	N	N	N	N	(0.03)	(0.17)	(0.07)	2.03	(0.05)	(0.3)	(13)	(0.31)	(2.0)	Tr

Composition of food per 100g

No.	Food	Description and main data sources	Water g	Total nitrogen g	Protein g	Fat g	Carbo-hydrate g	Energy value kcal	Energy value kJ
12-									
162	**Goats milk soft cheese**	4 samples, different types	65.0	2.06	13.1	15.8[a]	1.0	198	823
163	**Gouda**	10 samples	40.1	3.76	24.0	31.0	Tr	375	1555
164	**Gruyere**	10 samples	35.0	4.26	27.2	33.3	Tr	409	1695
165	**Hard cheese**, *average*	Average of Cheddar, Derby, Double Gloucester and Leicester	37.2	3.87	24.7	34.0	0.1	405	1679
166	**Lancashire**	10-22 samples	41.7	3.65	23.3	31.0	0.1	373	1545
167	**Leicester**	10-27 samples	37.5	3.81	24.3	33.7	0.1	401	1662
168	**Lymeswold**	Mild blue full fat soft cheese, 10 samples	41.0	2.44	15.6	40.3	Tr	425	1756
169	**Medium fat soft cheese**	e.g. Philadelphia Light, 5 samples, 3 brands	69.5	1.45	9.2	14.5	3.1	179	743
170	**Mozzarella**	10 samples	49.8	3.93	25.1	21.0	Tr	289	1204
171	**Parmesan**	10 samples, block and powdered	18.4	6.17	39.4	32.7	Tr	452	1880
172	**Processed cheese**, plain	10 samples, blocks and slices	45.7	3.26	20.8	27.0[b]	0.9	330	1367
173	smoked	10 assorted varieties	47.4	3.22	20.5	24.5	0.2	303	1258
174	**Quark**	Skimmed milk soft cheese (Magerquark), 10 samples	80.4	2.29	14.6	Tr	4.0	74	313
175	**Red Windsor**	10 samples	37.3	3.82	24.4	33.7	Tr	401	1662
176	**Ricotta**	Analysis and literature sources	72.1	1.48	9.4	11.0	2.0	144	599
177	**Roquefort**	10 samples	41.3	3.09	19.7	32.9	Tr	375	1552

[a] Some brands contain up to 25g fat per 100g

[b] Reduced fat varieties contain approximately 9.5g fat per 100g

Cheeses continued

Carbohydrate fractions and fatty acids, g per 100g
Cholesterol, mg per 100g

No. 12-	Food	Starch	Total sugars	Individual sugars						Fatty acids			Cholest- erol
				Gluc	Fruct	Galact	Sucr	Malt	Lact	Satd	Mono unsatd	Poly unsatd	
162	**Goats milk soft cheese**	0	1.0	0	0	0.1	0	0	0.9	10.4	3.6	0.5	N
163	**Gouda**	0	Tr	0	0	0	0	0	Tr	(19.4)	(9.0)	(0.9)	100
164	**Gruyere**	0	Tr	0	0	0	0	0	Tr	(20.8)	(9.7)	(1.0)	100
165	**Hard cheese**, average	0	0.1	0	0	0	0	0	0.1	21.3	9.9	1.0	100
166	**Lancashire**	0	0.1	0	0	0	0	0	0.1	19.4	9.0	0.9	90
167	**Leicester**	0	0.1	0	0	0	0	0	0.1	21.1	9.8	1.0	100
168	**Lymeswold**	0	Tr	0	0	0	0	0	Tr	25.2	11.7	1.2	115
169	**Medium fat soft cheese**	0	3.1	0	0	0	0	0	3.1	9.1	4.2	0.4	42
170	**Mozzarella**	0	Tr	0	0	0	0	0	Tr	(13.1)	(6.1)	(0.6)	65
171	**Parmesan**	0	Tr	0	0	0	0	0	Tr	(20.5)	(9.5)	(0.9)	100
172	**Processed cheese**, plain	0	0.9	0	0	0	0	0	0.9	16.6	7.7	1.2	85
173	smoked	0	0.2	0	0	0	0	0	0.2	15.1	7.0	1.1	80
174	**Quark**	0	4.0	0	0	Tr	0	0	4.0	Tr	Tr	Tr	1
175	**Red Windsor**	0	Tr	0	0	0	0	0	Tr	21.1	9.8	1.0	100
176	**Ricotta**	Tr	2.0	0	0	0	0	0	2.0	6.9	2.7	0.5	50
177	**Roquefort**	0	Tr	0	0	0	0	0	Tr	20.7	8.0	1.5	90

No. 12-	Food	Na	K	Ca	Mg	P	Fe	Cu	Zn	S	Cl	Mn	Se	I
						mg							µg	
162	Goats milk soft cheese	470	130	190	14	210	0.1	Tr	0.7	N	830	Tr	N	N
163	Gouda	910	91	740	38	490	0.1	Tr	1.8	N	1440	Tr	N	N
164	Gruyere	670	99	950	37	610	0.3	0.13	2.3	200	1040	Tr	N	N
165	Hard cheese, *average*	620	82	670	24	470	0.4	0.03	2.3	230	980	Tr	(12)	(44)
166	Lancashire	590	85	560	18	400	0.2	0.09	2.4	230	960	Tr	(11)	(46)
167	Leicester	630	85	660	22	470	0.5	0.06	3.3	230	910	Tr	(11)	(46)
168	Lymeswold	560	85	270	19	260	0.3	0.03	1.8	N	910	Tr	(7)	(46)
169	Medium fat soft cheese	N	N	N	N	N	N	N	N	N	N	Tr	(3)	N
170	Mozzarella	610	75	590	27	420	0.3	Tr	1.4	N	990	Tr	N	N
171	Parmesan	1090	110	1200	45	810	1.1	0.33	5.3	250	1820	0.1	11	N
172	Processed cheese, plain	1320	130	600	22	800	0.5	0.17	3.2	N	1100	Tr	(10)	(29)
173	smoked	1270	87	680	27	1030	0.3	0.20	3.2	N	820	Tr	(10)	(29)
174	Quark	45	140	120	11	200	Tr	0.06	0.9	110	110	Tr	N	4
175	Red Windsor	690	87	690	31	450	0.2	0.03	2.1	(230)	1030	Tr	(11)	(46)
176	Ricotta	100	110	240	13	170	0.4	N	1.3	N	N	Tr	N	N
177	Roquefort	1670	91	530	33	400	0.4	0.09	1.6	N	2670	Tr	N	N

Cheeses continued

No. 12-	Food	Retinol µg	Carotene µg	Vitamin D µg	Vitamin E mg	Thiamin mg	Ribo-flavin mg	Niacin mg	Trypt 60 mg	Vitamin B6 mg	Vitamin B12 µg	Folate µg	Panto-thenate mg	Biotin µg	Vitamin C mg
162	Goats milk soft cheese	310	(3)	(0.50)	0.79	0.04	0.63	0.65	3.01	0.12	2.0	19	N	N	Tr
163	Gouda	245	145	(0.24)	0.53	0.03	0.30	0.05	5.64	0.08	1.7	43	0.32	1.4	Tr
164	Gruyere	(325)	(225)	(0.25)	(0.58)	0.03	0.39	0.04	6.39	0.11	1.6	12	0.35	1.5	Tr
165	Hard cheese, average	335	225	0.26	0.50	0.03	0.43	0.07	5.81	0.11	1.3	28	0.34	3.0	Tr
166	Lancashire	325	215	0.24	0.71	0.03	0.45	0.11	5.47	0.08	1.1	44	0.27	4.0	Tr
167	Leicester	320	265	0.26	0.38	0.03	0.46	0.09	5.71	0.11	1.2	24	0.38	3.0	Tr
168	Lymeswold	440	330	0.31	0.93	0.04	0.43	0.56	3.66	0.14	0.9	56	0.39	6.3	Tr
169	Medium fat soft cheese	195	175	0.11	0.78	N	N	N	2.17	N	N	N	N	N	Tr
170	Mozzarella	240	170	(0.16)	0.33	0.03	0.31	0.08	5.89	0.09	2.1	19	0.25	2.2	Tr
171	Parmesan	345	210	(0.25)	0.70	0.03	0.44	0.12	9.25	0.13	1.9	12	0.43	3.3	Tr
172	Processed cheese, plain	270	95	0.21	0.55	0.03	0.28	0.10	4.89	0.08	0.9	18	0.31	2.3	Tr
173	smoked	(245)	(85)	0.19	(0.22)	0.03	0.27	0.06	4.83	0.07	0.9	18	0.22	1.5	Tr
174	Quark	2	1	Tr	Tr	0.04	0.30	0.19	3.43	0.08	0.7	45	0.44	3.0	1
175	Red Windsor	335	235	0.26	0.59	0.03	0.37	0.10	5.73	0.10	1.4	32	0.41	2.6	Tr
176	Ricotta	185	92	N	0.03	0.02	0.19	0.09	2.22	0.03	0.3	N	N	N	Tr
177	Roquefort	295	10	N	0.55	0.04	0.65	0.57	4.63	0.09	0.4	45	0.50	2.3	Tr

Cheeses continued

Composition of food per 100g

No.	Food	Description and main data sources	Water	Total nitrogen	Protein	Fat	Carbo-hydrate	Energy value	
12-			g	g	g	g	g	kcal	kJ
178	**Sage Derby**	10 samples, proximates as Derby	(38.0)	(3.79)	(24.2)	(33.9)	(0.1)	402	1667
179	**Soya cheese**	3 samples	47.7	2.93	18.3	27.3	Tr	319	1321
180	**Stilton**, blue	10-13 samples	38.6	3.56	22.7	35.5	0.1	411	1701
181	white	3-10 samples	45.8	3.12	19.9	31.3	0.1	362	1498
182	**Wensleydale**	10-23 samples	41.5	3.65	23.3	31.5	0.1	377	1563
183	**White cheese**, average	Average of Caerphilly, Cheshire, Lancashire, Wensleydale	41.4	3.67	23.4	31.3	0.1	376	1557

Carbohydrate fractions and fatty acids, g per 100g
Cholesterol, mg per 100g

No. 12-	Food	Starch	Total sugars	Individual sugars							Fatty acids			Cholest- erol
				Gluc	Fruct	Galact	Sucr	Malt	Lact	Satd	Mono unsatd	Poly unsatd	
178	**Sage Derby**	0	0.1	0	0	0	0	0	0.1	(21.2)	(9.8)	(1.0)	(100)
179	**Soya cheese**	Tr	Tr	0	0	0	0	0	0	N	N	N	Tr
180	**Stilton**, blue	0	0.1	0	0	0	0	0	0.1	22.2	10.3	1.0	105
181	white	0	0.1	0	0	0	0	0	0.1	19.6	9.1	0.9	90
182	**Wensleydale**	0	0.1	0	0	0	0	0	0.1	19.7	9.1	0.9	90
183	**White cheese**, *average*	0	0.1	0	0	0	0	0	0.1	19.6	9.1	0.9	90

Cheeses continued

Inorganic constituents per 100g

No. 12-	Food	Na	K	Ca	Mg	P	Fe	Cu	Zn	S	Cl	Mn	Se	I
							mg						µg	
178	**Sage Derby**	600	64	610	23	480	0.7	Tr	1.2	(230)	950	Tr	(11)	(46)
179	**Soya cheese**	600	130	450	29	350	1.1	Tr	1.8	N	870	0.1	N	N
180	**Stilton**, blue	930	130	320	20	310	0.3	0.18	2.5	230	1410	Tr	(11)	(46)
181	white	770	93	250	16	260	0.3	Tr	1.0	(230)	1310	Tr	(9)	(46)
182	**Wensleydale**	520	89	560	19	410	0.3	0.11	3.4	230	810	Tr	(11)	(46)
183	**White cheese**, average	530	88	560	19	400	0.4	0.12	3.1	230	840	Tr	(12)	(46)

No. 12-	Food	Retinol μg	Carotene μg	Vitamin D μg	Vitamin E mg	Thiamin mg	Ribo- flavin mg	Niacin mg	$\frac{Trypt}{60}$ mg	Vitamin B6 mg	Vitamin B12 μg	Folate μg	Panto- thenate mg	Biotin μg	Vitamin C mg
178	**Sage Derby**	345	215	0.26	0.53	0.03	0.43	0.05	5.69	0.09	1.4	24	0.34	3.2	Tr
179	**Soya cheese**	0	0[a]	0[a]	N	0.26	0.62	1.12	2.93	0.20	2.5	35	N	N	0
180	**Stilton**, blue	355	185	0.27	0.61	0.03	0.43	0.49	5.34	0.16	1.0	77	0.71	3.6	Tr
181	white	315	165	0.24	0.54	0.03	0.37	0.09	4.68	0.07	1.3	52	0.23	2.9	Tr
182	**Wensleydale**	275	260	0.24	0.39	0.03	0.46	0.11	5.47	0.09	1.1	43	0.30	4.0	Tr
183	**White cheese**, *average*	315	225	0.24	0.65	0.03	0.47	0.11	5.51	0.09	1.1	44	0.29	3.9	Tr

[a] β-Carotene and vitamin D may be added

71

Yogurts

12-184 to 12-197

Composition of food per 100g

No. 12-	Food	Description and main data sources	Water g	Total nitrogen g	Protein g	Fat g	Carbo-hydrate g	Energy value kcal	Energy value kJ
184	**Whole milk yogurt**, plain	22 samples, 2 brands	81.9	0.89	5.7	3.0	7.8	79	333
185	fruit	10 samples, assorted flavours, 'thick and creamy' type	73.1	0.80	5.1	2.8	15.7	105	441
186	'organic'	8 samples, 2 brands	87.0	0.67	4.3	2.9	5.8	56	236
187	goats	10 samples, 2 brands, from health food outlets	88.7	0.55	3.5	3.8	3.9	63	263
188	**Low fat yogurt**, plain	10 samples, 5 brands	84.9	0.80	5.1	0.8	7.5	56	236
189	flavoured	24 samples, 4 brands, assorted flavours	77.9	0.59	3.8	0.9	17.9	90	384
190	fruit	26 samples, 9 brands, assorted flavours	77.0	0.64	4.1	0.7	17.9	90	382
191	muesli/nut	20 samples, 10 hazelnut, 10 muesli	76.5	0.79	5.0	2.2	19.2	112	474
192	**Low calorie yogurt**	13 samples, 5 brands, assorted flavours with artificial sweeteners	87.9	0.67	4.3	0.2	6.0	41	177
193	**Drinking yogurt**	5 samples (Ambrosia), UHT	84.4	0.48	3.1	Tr[a]	13.1	62	263
194	**Greek yogurt**, cows	5 samples, 3 brands, 'strained' variety	78.5	1.01	6.4	9.1[b]	2.0	115	477
195	sheep	3 samples (Total), 'set' variety	80.9	0.69	4.4	7.5	5.6	106	442
196	**Soya yogurt**	5 samples sweetened (Sojal)	82.4	0.88	5.0	4.2	3.9	72	305
197	**Yogurt powder**	Commercial ingredient. Calculation and manufacturer's data	4.0	5.49	35.0	1.0	54.4	353	1502

[a] The fat content is variable. Non-UHT varieties contain 0.3 - 2g fat per 100g

[b] 'Set' varieties contain approximately 4g fat per 100g

Yogurts

Carbohydrate fractions and fatty acids, g per 100g
Cholesterol, mg per 100g

No. 12-	Food	Starch	Total sugars	Gluc	Fruct	Galact	Sucr	Malt	Lact	Satd	Mono unsatd	Poly unsatd	Cholesterol
						Individual sugars					Fatty acids		
184	**Whole milk yogurt**, plain	0	7.8	0	0	3.1	0	0	4.7	1.7	0.9	0.2	11
185	fruit	0	15.7[a]	2.2	1.8	1.4	6.3	0	4.0	1.5	0.8	0.2	10
186	'organic'	0	5.8	Tr	0	2.8	0	0	3.0	1.7	0.9	0.2	10
187	goats	0	3.9	0	0	2.6	0	0	1.3	2.5	0.9	0.1	11
188	**Low fat yogurt**, plain	0	7.5	0	0	2.9	0	0	4.6	0.5	0.2	Tr	4
189	flavoured	0	17.9	1.2	0.9	2.2	9.5	Tr	4.1	0.5	0.3	0.1	4
190	fruit	0	17.9	1.5	1.6	1.2	10.2	0	3.3	0.4	0.2	Tr	4
191	muesli/nut	1.2	17.9	1.8	1.4	3.2	7.6	0	4.0	N	N	N	(4)
192	**Low calorie yogurt**	0	6.0	0.2	0.4	1.3	0.1	0	4.0	0.1	0.1	Tr	1
193	**Drinking yogurt**	0	13.1	1.3	1.1	1.9	6.0	0	2.8	Tr	Tr	Tr	Tr
194	**Greek yogurt**, cows	0	2.0	0.1	0.2	1.1	0.1	0	0.5	5.2	2.7	0.5	N
195	sheep	0	5.6	0	0	1.5	0	0	4.1	4.8	1.9	0.4	(14)
196	**Soya yogurt**	Tr	3.9	1.2	1.4	0	1.3	0	0	0.6	0.9	2.4	0
197	**Yogurt powder**	0	54.4	0	0	(3.7)	0	0	(50.7)	0.6	0.3	Tr	22

[a] 'Real' fruit yogurts contain 12.1g total sugars as 1.2g glucose, 5.4g fructose, 1.5g galactose, 0.2g sucrose and 3.8g lactose per 100g

Inorganic constituents per 100g

No. 12-	Food	Na	K	Ca	Mg	P	Fe	Cu	Zn	S	Cl	Mn	Se	I
						mg							µg	
184	**Whole milk yogurt**, plain	80	280	200	19	170	0.1	Tr	0.7	53	170	Tr	(2)	(63)
185	fruit	82	210	160	16	130	Tr	Tr	0.5	23	150	Tr	(1)	(48)
186	'organic'	65	150	140	13	110	0.1	Tr	0.4	40	130	Tr	(1)	N
187	goats	39	170	120	14	110	0.2	0.01	0.4	N	130	Tr	N	N
188	**Low fat yogurt**, plain	83	250	190	19	160	0.1	Tr	0.6	48	150	Tr	1	63
189	flavoured	65	190	150	15	120	0.1	Tr	0.5	36	130	Tr	(1)	N
190	fruit	64	210	150	15	120	0.1	Tr	0.5	38	130	Tr	(1)	48
191	muesli/nut	80	240	170	22	150	0.2	N	0.7	N	130	N	N	N
192	**Low calorie yogurt**	73	180	130	13	110	0.1	Tr	0.4	40	120	Tr	(1)	N
193	**Drinking yogurt**	47	130	100	11	81	0.1	0.01	0.3	29	75	Tr	(1)	N
194	**Greek yogurt**, cows	71	150	150	12	130	0.3	Tr	0.5	N	100	Tr	2	N
195	sheep	150	190	150	16	140	Tr	Tr	0.5	N	220	Tr	1	N
196	**Soya yogurt**	N	N	N	N	N	N	N	N	N	N	N	N	N
197	**Yogurt powder**	590	1620	1300	130	1030	0.5	0.03	4.3	330	1080	Tr	(11)	(160)

No. 12-	Food	Retinol µg	Carotene µg	Vitamin D µg	Vitamin E mg	Thiamin mg	Ribo-flavin mg	Niacin mg	Trypt 60 mg	Vitamin B6 mg	Vitamin B12 µg	Folate µg	Panto-thenate mg	Biotin µg	Vitamin C mg
184	**Whole milk yogurt**, plain	28	21	0.04	0.05	0.06	0.27	0.18	1.33	0.10	0.2	18	0.50	2.6	1
185	fruit	39	16	(0.04)	(0.05)	0.06	0.30	0.13	1.29	0.07	0.1	10	0.30	2.0	1
186	'organic'	27	20	0.04	0.05	0.04	0.20	0.09	1.01	0.05	0.1	11	0.25	1.4	1
187	goats	N	Tr	N	0.03	0.04	0.17	0.27	0.83	0.06	Tr	7	0.23	0.5	1
188	**Low fat yogurt**, plain	8	5	0.01	0.01	0.05	0.25	0.15	1.20	0.09	0.2	17	0.45	2.9	1
189	flavoured	9	6	0.01	0.01	0.05	0.21	0.12	0.89	0.07	0.2	19	0.30	2.2	1
190	fruit	10	4	(0.01)	(0.01)	0.05	0.21	0.14	0.96	0.08	0.2	16	0.33	2.3	1
191	muesli/nut	(8)	(23)	(0.03)	(0.15)	0.07	0.24	0.19	1.19	0.10	0.2	10	0.38	2.9	1
192	**Low calorie yogurt**	Tr	Tr	Tr	0.03	0.04	0.29	0.13	1.00	0.07	(0.2)	8	N	N	1
193	**Drinking yogurt**	Tr	Tr	Tr	Tr	0.03	0.16	0.09	0.72	0.05	0.2	12	0.19	0.9	0
194	**Greek yogurt**, cows	115	(36)	0.05	0.38	0.03	0.36	0.06	1.51	0.05	0.2	6	N	N	Tr
195	sheep	86	(11)	0.24	0.73	0.05	0.33	0.23	1.03	0.08	0.2	3	N	N	Tr
196	**Soya yogurt**	23	(3)	0	1.49	N	N	N	0.88	N	0	N	N	N	0
197	**Yogurt powder**	13	5	0.01	0.02	N	N	N	8.23	N	N	N	N	N	N

No. 12-	Food	Description and main data sources	Water g	Total nitrogen g	Protein g	Fat g	Carbohydrate g	Energy value kcal	kJ
198	**Arctic roll**	10 samples, 2 brands	51.3	0.66	4.1	6.6	33.3	200	847
199	**Banana split**	Recipe	64.8	0.40	2.4	11.2	19.3ᵃ	182	761
200	**Choc ice**	Plain and milk varieties; analysis and manufacturer's data (Birds Eye Wall's)	N	0.55	3.5	17.5	28.1	277	1157
201	**Chocolate nut sundae**	Recipe	46.0	0.49	3.0	15.3	34.2ᵃ	278	1165
202	**Cornetto**	Analysis and manufacturer's data (Birds Eye Wall's)	N	0.59	3.7	12.9	34.5	260	1092
203	**Frozen ice cream desserts**	6 samples, different types eg Sonata, Viennetta	61.7	0.53	3.3	14.2	22.8	227	946
204	**Ice cream**, dairy, vanilla	17 samples	61.9	0.56	3.6	9.8	24.4ᵃ	194	814
205	flavoured	17 samples, assorted flavours	59.8	0.54	3.5	8.0	24.7ᵃ	179	751
206	non-dairy, vanilla	11 samples, hard and soft scoop	65.3	0.50	3.2	8.7	23.1ᵃ	178	746
207	flavoured	14 samples, hard and soft scoop assorted flavours	64.9	0.49	3.1	7.4	23.2ᵃ	166	698
208	mixes	Prepared mix from ice cream parlour	63.4	0.65	4.1	7.9	25.1ᵃ	182	764
209	reduced calorie	2 samples (Heinz)	64.1	0.53	3.4	6.0	13.7ᵇ	119	499
210	*with cone*	Recipe	62.6	0.56	3.5	8.5	25.5ᵃ	186	781
211	*with wafers*	Recipe	62.9	0.55	3.5	8.5	25.2ᵃ	185	777
212	**Ice cream wafers**	6 samples, 2 brands	2.8	1.77	10.1	3.3	78.8	366	1555

ᵃ Including oligosaccharides from the glucose syrup/maltodextrins in the product

ᵇ Also contains about 12g polydextrose

Carbohydrate fractions and fatty acids, g per 100g
Cholesterol, mg per 100g

No. 12-	Food	Starch	Total sugars	Individual sugars						Fatty acids			Cholest-erol
				Gluc	Fruct	Galact	Sucr	Malt	Lact	Satd	Mono unsatd	Poly unsatd	
198	**Arctic roll**	8.0	25.3	2.6	1.1	0.1	16.9	1.7	2.9	3.1	2.5	0.8	30
199	**Banana split**	N	16.3[a]	3.2	1.9	0	7.7	0.4	3.1	6.0	3.6	1.0	19
200	**Choc ice**	N	N	N	N	0	N	N	N	10.8	4.8	1.1	7
201	**Chocolate nut sundae**	0.4	31.9[a]	(1.9)	(0.9)	0	(22.5)	(0.7)	(5.8)	8.3	4.9	1.2	28
202	**Cornetto**	9.0	25.5	0.7	0.9	0	19.0	0.6	4.3	6.7[b]	4.2[b]	1.3[b]	2[b]
203	**Frozen ice cream desserts**	Tr	22.8	1.1	0.1	0	16.1	0.6	4.9	11.2	1.8	0.4	3
204	**Ice cream**, dairy, vanilla	Tr	22.1[a]	2.4	0	0	13.9	Tr	5.9	6.4	2.4	0.3	31
205	flavoured	Tr	23.7[a]	5.2	1.5	0	12.3	Tr	4.7	5.2	2.0	0.3	26
206	non-dairy, vanilla	Tr	19.2[a]	1.0	0.1	0	11.5	Tr	6.5	4.4	3.2	0.8	7
207	flavoured	Tr	21.3[a]	2.2	0.6	0	12.7	Tr	5.8	3.7	2.7	0.6	6
208	mixes	Tr	21.7[a]	Tr	0	0	15.2	Tr	6.5	4.0	2.9	0.7	(7)
209	reduced calorie	0	13.7	0.1	8.4	0	0	0	5.2	N	N	N	N
210	with cone	3.4	18.4[a]	1.0	0.1	0	11.0	Tr	6.2	N	N	N	6
211	with wafers	3.0	18.5[a]	1.0	0.1	0	11.1	Tr	6.2	N	N	N	6
212	**Ice cream wafers**	77.7	1.1	0.1	0.1	0	0.7	0.2	0	N	N	N	0

[a] Not including oligosaccharides from the glucose syrup/maltodextrins in the product

[b] Strawberry variety only

Ice creams

Inorganic constituents per 100g

No. 12-	Food	Na	K	Ca	Mg	P	Fe	Cu	Zn	S	Cl	Mn	Se	I
							mg						µg	
198	**Arctic roll**	150	140	90	11	120	0.7	0.12	0.4	65	140	0.1	N	23
199	**Banana split**	37	260	62	25	69	0.3	0.06	0.3	N	95	0.2	N	N
200	**Choc ice**	91	200	130	27	N	0.1	Tr	0.2	N	N	N	N	N
201	**Chocolate nut sundae**	150	170	80	N	N	N	N	N	N	N	N	N	N
202	**Cornetto**	91	170	120	21	N	N	N	N	N	N	N	N	N
203	**Frozen ice cream desserts**	84	200	110	19	99	0.5	0.04	0.4	N	110	0.1	N	N
204	**Ice cream**, dairy, vanilla	69	160	130	13	110	0.1	0.02	0.3	N	110	Tr	N	N
205	flavoured	61	180	110	19	99	0.5	0.05	0.4	N	100	Tr	N	N
206	non-dairy, vanilla	76	170	120	13	100	0.1	Tr	0.3	N	130	Tr	N	N
207	flavoured	72	160	120	13	99	0.1	Tr	0.4	N	120	Tr	N	N
208	mixes	59	180	140	15	120	0.1	Tr	0.5	N	110	Tr	N	N
209	reduced calorie	N	N	N	N	N	N	N	N	N	N	N	N	N
210	with cone	76	170	120	14	100	0.2	Tr	0.3	N	130	Tr	N	N
211	with wafers	76	170	120	14	100	0.2	Tr	0.3	N	130	Tr	N	N
212	**Ice cream wafers**	93	190	170	46	130	2.0	0.11	0.7	N	130	0.7	N	N

No. 12-	Food	Retinol µg	Carotene µg	Vitamin D µg	Vitamin E mg	Thiamin mg	Ribo-flavin mg	Niacin mg	Trypt 60 mg	Vitamin B6 mg	Vitamin B12 µg	Folate µg	Panto-thenate mg	Biotin µg	Vitamin C mg
198	**Arctic roll**	N	N	N	N	0.07	0.11	0.24	0.88	0.06	0.3	12	N	N	0
199	**Banana split**	77	53	0.03	0.77	0.05	0.15	0.50	0.55	0.16	0.2	11	0.39	2.5	5
200	**Choc ice**	1	5	Tr	N	N	N	N	0.83	N	N	N	N	N	N
201	**Chocolate nut sundae**	115	65	0.05	N	0.05	0.15	0.29	0.60	0.05	0.3	7	0.29	1.7	1
202	**Cornetto**	N	N	Tr	N	N	N	N	N	N	N	N	N	N	N
203	**Frozen ice cream desserts**	2	5	Tr	0.51	0.04	0.30	0.17	0.79	0.06	0.6	3	N	N	0
204	**Ice cream,** dairy, vanilla	115	195	0.12	0.21	0.04	0.25	0.13	0.84	0.08	0.4	7	0.44	2.5	1
205	flavoured	94	160	0.10	0.17	0.04	0.26	0.17	0.81	0.07	0.3	9	0.33	2.4	1
206	non-dairy, vanilla	1	6	Tr	0.84	0.04	0.24	0.12	0.75	0.07	0.5	8	0.43	3.0	1
207	flavoured	1	5	Tr	0.72	0.04	0.24	0.13	0.73	0.07	0.4	8	0.40	2.8	1
208	mixes	2	4	Tr	0.68	0.05	0.23	0.11	0.97	0.08	0.3	6	0.44	3.1	1
209	reduced calorie	N	N	Tr	N	N	N	N	0.79	N	N	N	N	N	N
210	with cone	Tr	5	Tr	0.77	0.05	0.23	0.21	0.81	0.07	0.5	8	N	N	1
211	with wafers	Tr	5	Tr	0.78	0.04	0.23	0.18	0.78	0.07	0.5	8	N	N	1
212	**Ice cream wafers**	0	0	0	N	0.20	0.04	2.30	2.07	0.15	0	15	N	N	0

Ice creams *continued*

Composition of food per 100g

No.	Food	Description and main data sources	Water	Total nitrogen	Protein	Fat	Carbo-hydrate	Energy value	
12-			g	g	g	g	g	kcal	kJ
213	**Knickerbocker glory**	Recipe	76.9	0.25	1.5	5.0	16.4[a]	112	473
214	**Kulfi**	Indian ice cream. Recipe	41.1	0.89	5.4	39.9	11.8	424	1756
215	**Peach melba**	Recipe	70.5	0.31	2.0	10.7	15.9[a]	164	685
216	**Sorbet**, lemon	Recipe	64.9	0.14	0.9	Tr	34.2	131	562

[a] Including oligosaccharides from the glucose syrup/maltodextrins in the product

Ice creams *continued*

Carbohydrate fractions and fatty acids, g per 100g
Cholesterol, mg per 100g

No. 12-	Food	Starch	Total sugars	Individual sugars						Fatty acids			Cholest-erol
				Gluc	Fruct	Galact	Sucr	Malt	Lact	Satd	Mono unsatd	Poly unsatd	
213	**Knickerbocker glory**	Tr	15.5[a]	3.1	1.9	0	8.0	1.0	1.6	2.9	1.6	0.3	10
214	**Kulfi**	0	11.8	2.4	2.8	0	0.3	0	6.3	22.8	13.1	1.8	100
215	**Peach melba**	0.6	13.7[a]	1.9	1.7	0	6.9	Tr	3.2	6.3	3.4	0.5	22
216	**Sorbet**, lemon	0	34.2	0.1	0.2	0	33.9	0	0	Tr	Tr	Tr	0

[a] Not including oligosaccharides from the glucose syrup/maltodextrins in the product

Ice creams *continued*

Inorganic constituents per 100g

No. 12-	Food	Na	K	Ca	Mg	P	Fe	Cu	Zn	S	Cl	Mn	Se	I
							mg						μg	
213	**Knickerbocker glory**	22	63	34	5	27	0.3	0.01	0.1	6	36	0.1	N	N
214	**Kulfi**	76	240	160	33	150	0.5	0.01	0.7	49	130	0.1	1	15
215	**Peach melba**	42	140	63	9	58	0.2	0.01	0.2	4	66	Tr	N	N
216	**Sorbet**, lemon	18	42	2	2	5	Tr	0.03	1.0	17	16	Tr	1	Tr

No. 12-	Food	Retinol µg	Carotene µg	Vitamin D µg	Vitamin E mg	Thiamin mg	Ribo-flavin mg	Niacin mg	Trypt 60 mg	Vitamin B6 mg	Vitamin B12 µg	Folate µg	Panto-thenate mg	Biotin µg	Vitamin C mg
213	**Knickerbocker glory**	39	30	0.02	N	0.02	0.06	0.10	0.20	0.03	0.1	3	0.12	0.7	2
214	**Kulfi**	450	240	0.21	N	0	N	N	1.17	N	N	N	N	N	0
215	**Peach melba**	105	79	0.04	N	0.03	0.13	0.30	0.43	0.05	0.2	6	0.24	1.5	3
216	**Sorbet**, lemon	0	0	0	Tr	Tr	0.04	Tr	0.25	0.01	Tr	1	0.04	0.7	Tr

Puddings and chilled desserts

Composition of food per 100g

No. 12-	Food	Description and main data sources	Water g	Total nitrogen g	Protein g	Fat g	Carbo-hydrate g	Energy value kcal	kJ
217	**Blancmange**	Recipe	75.5	0.48	3.1	3.7	18.2	114	480
218	**Cheesecake**	Recipe	34.4	0.61	3.7	35.5	24.6	426	1769
219	*frozen*	10 samples, assorted flavours, fruit topping	44.0	0.91	5.7	10.6	33.0	242	1017
220	**Creme caramel**	9 samples, 4 brands	72.0	0.47	3.0	2.2	20.6	109	462
221	*homemade*	Recipe	62.6	0.78	4.9	4.8	28.3	169	714
222	**Custard,** *made up with* *whole milk*	Recipe	75.5	0.58	3.7	4.5	16.6	117	495
223	*made up with* *semi-skimmed milk*	Recipe	77.8	0.60	3.8	1.9	16.8	94	403
224	*made up with skimmed milk*	Recipe	79.3	0.60	3.8	0.1	16.8	79	339
225	*canned*	10 samples, 3 brands	77.2	0.42	2.6	3.0	15.4	95	401
226	*confectioners'*	Recipe	63.5	1.03	6.4	5.9	24.4	170	718
227	*egg*	Recipe	77.0	0.90	5.7	6.0	11.0	118	494
228	**Dream Topping**	10 samples (Birds)	1.4	0.95	6.0	50.4	39.8	626	2603
229	*made up with whole milk*	Recipe	69.9	0.59	3.8	13.5	12.1	182	757
230	*made up with* *semi-skimmed milk*	Recipe	71.5	0.61	3.9	11.7	12.2	166	694
231	*made up with skimmed milk*	Recipe	72.5	0.61	3.9	10.5	12.2	155	650
232	**Fruit fool**	Recipe	66.6	0.16	1.0	9.3	20.2	163	683

84

Puddings and chilled desserts

Carbohydrate fractions and fatty acids, g per 100g
Cholesterol, mg per 100g

No. 12-	Food	Starch	Total sugars	Individual sugars						Fatty acids			Cholest-erol
				Gluc	Fruct	Galact	Sucr	Malt	Lact	Satd	Mono unsatd	Poly unsatd	
217	**Blancmange**	6.4	11.8	Tr	0	0	7.3	0	4.5	2.3	1.1	0.1	13
218	**Cheesecake**	11.0	13.6	0.1	0.1	0	12.9	0	0.5	18.8	11.4	2.8	110
219	*frozen*	10.8	22.2	2.7	1.4	Tr	16.0	0.4	1.7	5.6	3.6	0.8	60
220	**Creme caramel**	2.6	18.0	2.3	1.3	0.1	10.3	0.5	3.5	N	N	N	N
221	*homemade*	0	28.3	Tr	0	0	25.8	0	2.5	2.1	1.8	0.4	105
222	**Custard**, *made up with whole milk*	5.1	11.4	Tr	0	0	5.9	0	5.5	2.8	1.3	0.2	16
223	*made up with semi-skimmed milk*	5.1	11.6	Tr	0	0	5.9	0	5.8	1.2	0.5	0.1	8
224	*made up with skimmed milk*	5.1	11.6	Tr	0	0	5.9	0	5.8	0.1	Tr	Tr	2
225	*canned*	3.1	12.3	Tr	Tr	0	7.9	0.3	4.0	1.7	0.9	0.1	(11)
226	*confectioners'*	5.5	18.9	Tr	Tr	0	15.1	Tr	3.8	2.7	2.1	0.5	110
227	*egg*	0	11.0	Tr	0	0	6.2	0	4.8	3.1	2.1	0.4	90
228	**Dream Topping**	9.9	29.9	Tr	Tr	0.2	21.5	1.2	7.0	46.9	0.6	0.4	4
229	*made up with whole milk*	2.1	10.0	Tr	Tr	Tr	4.5	0.2	5.3	11.7	1.0	0.2	11
230	*made up with semi-skimmed milk*	2.1	10.2	Tr	Tr	Tr	4.5	0.2	5.4	10.5	0.5	0.1	6
231	*made up with skimmed milk*	2.1	10.2	Tr	Tr	Tr	4.5	0.2	5.4	9.8	0.1	0.1	2
232	**Fruit fool**	4.1	16.1	0.8	0.9	0	13.4	0.2	0.7	5.8	2.7	0.3	24

85

Puddings and chilled desserts

Inorganic constituents per 100g

No. 12-	Food	Na	K	Ca	Mg	P	mg Fe	Cu	Zn	S	Cl	Mn	µg Se	I
217	**Blancmange**	55	140	110	10	89	0.1	0.01	0.4	28	99	0	1	14
218	**Cheesecake**	290	120	65	10	81	0.8	0.08	0.4	N	380	0.1	N	N
219	*frozen*	160	130	68	10	93	0.5	0.06	0.4	55	220	0.2	N	25
220	**Creme caramel**	70	150	94	9	77	Tr	Tr	0.3	N	100	Tr	N	33
221	homemade	64	110	74	8	99	0.5	0.03	0.6	62	93	Tr	3	21
222	**Custard**, *made up with whole milk*	81	160	130	13	110	0.1	Tr	0.4	34	140	N	N	N
223	*made up with semi-skimmed milk*	81	180	140	13	110	0.1	Tr	0.5	35	140	N	N	N
224	*made up with skimmed milk*	81	180	140	14	110	0.1	Tr	0.5	35	140	N	N	N
225	*canned*	67	130	100	8	87	0.2	0.02	0.3	32	75	Tr	N	N
226	*confectioners'*	79	150	110	13	130	0.7	0.03	0.7	69	130	0.1	4	26
227	*egg*	82	170	130	13	130	0.4	0.02	0.6	65	130	Tr	3	25
228	**Dream Topping**	130	63	23	6	95	0.5	0.13	0.3	78	15	Tr	N	Tr
229	*made up with whole milk*	70	120	95	9	92	0.1	0.03	0.3	39	82	Tr	N	12
230	*made up with semi-skimmed milk*	70	130	99	9	94	0.1	0.03	0.4	40	82	Tr	N	12
231	*made up with skimmed milk*	70	130	99	10	94	0.1	0.03	0.4	40	82	Tr	N	12
232	**Fruit fool**	13	160	38	8	31	0.4	0.06	0.2	13	32	0.2	N	N

Puddings and chilled desserts

No. 12-	Food	Retinol µg	Carotene µg	Vitamin D µg	Vitamin E mg	Thiamin mg	Ribo-flavin mg	Niacin mg	Trypt 60 mg	Vitamin B6 mg	Vitamin B12 µg	Folate µg	Panto-thenate mg	Biotin µg	Vitamin C mg
217	**Blancmange**	49	19	0.03	0.07	0.03	0.14	0.1	0.7	0.05	0.4	4	0.30	1.8	Tr
218	**Cheesecake**	315	190	0.94	N	0.04	0.12	0.2	0.9	0.04	0.5	6	N	2.9	1
219	*frozen*	N	N	N	N	0.04	0.16	0.3	1.4	0.02	0.5	7	N	N	0
220	**Creme caramel**	37	8	0.07	0.16	0.03	0.20	0.1	0.7	0.03	0.3	8	N	N	0
221	*homemade*	76	10	0.47	0.27	0.03	0.18	0.1	1.3	0.05	0.8	8	0.49	6.2	Tr
222	**Custard,** *made up with whole milk*	59	24	0.03	0.10	0.04	0.18	0.1	0.9	0.06	0.5	5	0.36	2.2	1
223	*made up with semi-skimmed milk*	24	10	0.01	0.03	0.04	0.19	0.1	0.9	0.06	0.5	5	0.33	2.3	1
224	*made up with skimmed milk*	1	Tr	Tr	Tr	0.04	0.19	0.1	0.9	0.06	0.5	5	0.33	2.3	1
225	*canned*	N	N	N	N	0.04	0.10	Tr	0.6	0.03	Tr	2	N	N	0
226	*confectioners'*	89	16	0.47	0.37	0.07	0.23	0.2	1.7	0.08	1.0	15	0.68	6.7	1
227	*egg*	89	21	0.37	0.24	0.05	0.24	0.1	1.5	0.08	0.8	12	0.63	5.8	Tr
228	**Dream Topping**	Tr	N	0	N	0.02	0.25	Tr	1.1	0.02	0.8	Tr	N	N	0
229	*made up with whole milk*	41	N	0.02	N	0.04	0.19	0.1	0.8	0.05	0.5	4	N	N	1
230	*made up with semi-skimmed milk*	16	N	0.01	N	0.04	0.19	0.1	0.9	0.05	0.5	4	N	N	1
231	*made up with skimmed milk*	Tr	N	Tr	N	0.04	0.19	0.1	0.9	0.05	0.5	4	N	N	1
232	**Fruit fool**	130	120	0.05	0.45	0.02	0.06	0.2	0.2	0.03	Tr	N	0.14	0.9	17

Puddings and chilled desserts *continued*

Composition of food per 100g

No. 12-	Food	Description and main data sources	Water g	Total nitrogen g	Protein g	Fat g	Carbo-hydrate g	Energy value kcal	kJ
233	**Instant dessert powder**	10 samples, 2 types, assorted flavours	1.0	0.39	2.4	17.3	60.1	391	1643
234	*made up with whole milk*	Recipe	72.1	0.48	3.1	6.3	14.8	111	467
235	*made up with*								
	semi-skimmed milk	Recipe	73.8	0.50	3.1	4.4	14.9	94	401
236	*made up with skimmed milk*	Recipe	74.9	0.50	3.1	3.2	14.9	84	356
237	**Jelly**, *made with water*	Recipe	84.0	0.21	1.2	0	15.1	61	260
238	*made with whole milk*	Recipe	79.1	0.42	2.6	1.7	16.7	88	373
239	*made with semi-skimmed milk*	Recipe	80.0	0.43	2.6	0.7	16.8	79	338
240	*made with skimmed milk*	Recipe	80.6	0.43	2.6	Tr	16.8	73	314
241	**Milk pudding,** *made*								
	with whole milk	e.g. rice, sago, semolina, tapioca; recipe	72.4	0.62	3.9	4.3	19.9	129	543
242	*made with semi-skimmed milk*	e.g. rice, sago, semolina, tapioca; recipe	74.6	0.64	4.0	1.8	20.1	107	457
243	*made with skimmed milk*	e.g. rice, sago, semolina, tapioca; recipe	76.0	0.64	4.0	0.2	20.1	93	398
244	**Mousse**, chocolate	10 samples, 4 brands, fresh	67.3	0.63	4.0	5.4	19.9	139	586
245	chocolate, rich	6 samples, Chambourcy and own brand	60.0	0.74	4.7	6.9	26.0	178	751
246	fruit	8 samples, assorted flavours, fresh	71.7	0.71	4.5	5.7	18.0	137	575
247	frozen	10 samples, 7 brands, assorted flavours	64.7	0.58	3.7	7.0	21.0	157	658
248	**Rice pudding,** canned	10 cans, 4 brands	77.6	0.53	3.4	2.5	14.0	89	374
249	**Trifle**	Recipe	67.2	0.59	3.6	6.3	22.3	160	674
250	*frozen*	10 samples, 5 brands, strawberry and raspberry	67.7	0.35	2.2	5.8	20.6	138	581
251	*with Dream Topping*	Recipe	68.2	0.61	3.7	4.8	22.7	148	624
252	*with fresh cream*	10 samples, individual and large	68.1	0.38	2.4	9.2	19.5	166	693

Puddings and chilled desserts *continued*

Carbohydrate fractions and fatty acids, g per 100g
Cholesterol, mg per 100g

| No. 12- | Food | Starch | Total sugars | Individual sugars | | | | | | Fatty acids | | | Cholest-erol |
				Gluc	Fruct	Galact	Sucr	Malt	Lact	Satd	Mono unsatd	Poly unsatd	
233	**Instant dessert powder**	19.4	40.7	Tr	Tr	0.2	38.3	0	2.2	15.9	0.3	0.2	1
234	*made up with whole milk*	3.5	11.3	Tr	Tr	Tr	6.9	0	4.3	4.9	1.0	0.1	12
235	*made up with*												
	semi-skimmed milk	3.5	11.4	Tr	Tr	Tr	6.9	0	4.5	3.7	0.4	0.1	5
236	*made up with skimmed milk*	3.5	11.4	Tr	Tr	Tr	6.9	0	4.5	2.9	0.1	Tr	2
237	**Jelly**, *made with water*	0	15.1	3.5	1.7	0	8.6	1.3	0	0	0	0	0
238	*made with whole milk*	0	16.7	3.4	1.7	0	8.3	1.3	2.1	1.1	0.5	0.1	6
239	*made with semi-skimmed milk*	0	16.8	3.4	1.7	0	8.3	1.3	2.2	0.4	0.2	Tr	3
240	*made with skimmed milk*	0	16.8	3.4	1.7	0	8.3	1.3	2.2	Tr	Tr	Tr	Tr
241	**Milk pudding**, *made*												
	with whole milk	9.3	10.7	Tr	Tr	0	5.5	0	5.2	2.7	1.2	0.2	15
242	*made with semi-skimmed milk*	9.3	10.9	Tr	Tr	0	5.5	0	5.4	1.1	0.5	0.1	7
243	*made with skimmed milk*	9.3	10.9	Tr	Tr	0	5.5	0	5.4	0.1	Tr	Tr	2
244	**Mousse**, chocolate	2.4	17.5	1.1	1.8	0	10.8	0	3.8	N	N	N	N
245	chocolate, rich	0.7	25.3	0.1	0.1	0	20.9	0	4.2	N	N	N	N
246	fruit	Tr	18.0	3.1	2.9	0.4	7.6	0.3	3.7	N	N	N	N
247	frozen	1.8	19.3	1.4	0.5	Tr	12.0	2.0	3.3	3.7	2.3	0.6	4
248	**Rice pudding**, canned	5.8	8.2a	Tr	0	0	5.0	0	3.2	1.6	0.7	0.1	(9)
249	**Trifle**	5.5	16.8	2.6	2.0	0	8.8	0.8	2.7	3.1	2.0	0.7	44
250	frozen	2.7	17.9	4.8	0.6	0	10.1	1.0	1.4	3.2	1.9	0.3	20
251	with Dream Topping	5.6	17.1	2.6	2.0	0	9.0	0.8	2.8	2.3	1.5	0.6	38
252	with fresh cream	4.5	15.0	1.7	1.4	0	9.6	0.5	1.8	5.2	3.0	0.5	33

a Low calorie varieties contain approximately 3.1g sugar per 100g

Puddings and chilled desserts *continued*

No. 12-	Food	Na	K	Ca	Mg	P	mg Fe	Cu	Zn	S	Cl	Mn	µg Se	I
233	**Instant dessert powder**	1100	64	20	11	650	0.5	0.20	0.4	47	45	0.1	N	Tr
234	*made up with whole milk*	240	130	97	10	190	0.1	0.04	0.4	33	90	Tr	N	12
235	*made up with semi-skimmed milk*	240	130	100	10	190	0.1	0.04	0.4	33	90	Tr	N	12
236	*made up with skimmed milk*	240	130	100	11	190	0.1	0.04	0.4	33	90	Tr	N	12
237	**Jelly**, *made with water*	5	5	7	Tr	1	0.4	0.01	N	8	6	N	N	N
238	*made with whole milk*	29	66	57	5	41	0.4	0.01	N	21	50	N	N	N
239	*made with semi-skimmed milk*	29	71	59	5	43	0.4	0.01	Tr	21	50	N	N	N
240	*made with skimmed milk*	29	71	59	6	43	0.4	0.01	N	21	50	N	N	N
241	**Milk pudding**, *made with whole milk*	59	160	130	13	110	0.1	0.01	0.5	36	110	N	N	N
242	*made with semi-skimmed milk*	59	170	130	13	110	0.1	0.01	0.5	38	110	N	N	N
243	*made with skimmed milk*	59	170	130	14	110	0.1	0.01	0.5	38	110	N	N	N
244	**Mousse**, chocolate	67	220	97	28	100	1.6	0.12	0.6	N	86	0.2	N	N
245	chocolate, rich	N	N	N	N	N	N	N	N	N	N	N	N	N
246	fruit	62	150	120	12	96	Tr	Tr	0.4	N	110	Tr	N	N
247	frozen	57	130	93	11	69	0.2	0.04	0.3	40	110	Tr	N	N
248	**Rice pudding**, canned	50	140	93	11	80	0.2	0.03	0.4	N	95	N	N	N
249	**Trifle**	53	140	79	15	85	0.5	0.04	0.4	N	87	0.1	N	83
250	frozen	80	29	49	5	52	0.2	0.04	0.2	30	45	Tr	N	(17)
251	with Dream Topping	55	150	82	15	87	0.5	0.04	0.4	N	88	0.1	N	84
252	with fresh cream	63	84	68	6	63	0.3	0.09	0.3	26	55	0.1	N	17

Puddings and chilled desserts *continued*

No. 12-	Food	Retinol µg	Carotene µg	Vitamin D µg	Vitamin E mg	Thiamin mg	Ribo-flavin mg	Niacin mg	Trypt 60 mg	Vitamin B6 mg	Vitamin B12 µg	Folate µg	Panto-thenate mg	Biotin µg	Vitamin C mg
233	**Instant dessert powder**	N	N	N	N	Tr	0.01	Tr	0.5	Tr	0.3	Tr	N	N	0
234	*made up with whole milk*	N	N	N	N	0.03	0.14	0.1	0.7	0.05	0.3	4	N	N	1
235	*made up with semi-skimmed milk*	N	N	N	N	0.03	0.15	0.1	0.7	0.05	0.3	4	N	N	1
236	*made up with skimmed milk*	N	N	N	N	0.03	0.15	0.1	0.7	0.05	0.3	4	N	N	1
237	**Jelly**, *made with water*	0	0	0	0	0	0	0	0	0	0	0	0	0	0
238	*made with whole milk*	22	9	0.01	0.04	0.02	0.07	Tr	0.3	0.03	Tr	2	0.15	0.8	Tr
239	*made with semi-skimmed milk*	9	3	Tr	0.01	0.02	0.08	Tr	0.3	0.03	0.2	2	0.14	0.9	1
240	*made with skimmed milk*	Tr	Tr	Tr	Tr	0.02	0.08	Tr	0.3	0.03	0.2	2	0.14	0.9	1
241	**Milk pudding**, *made with whole milk*	56	22	0.03	0.10	0.04	0.16	0.1	0.9	0.06	0.4	3	0.30	2.2	1
242	*made with semi-skimmed milk*	22	9	0.01	0.03	0.04	0.17	0.1	0.9	0.06	0.4	3	0.28	2.3	1
243	*made with skimmed milk*	1	Tr	Tr	Tr	0.04	0.17	0.1	0.9	0.06	0.4	3	0.28	2.3	1
244	**Mousse**, chocolate	46	11	Tr	0.58	0.04	0.21	0.2	0.9	0.04	0.2	6	N	N	0
245	chocolate, rich	N	N	Tr	N	N	N	N	1.1	N	N	N	N	N	0
246	fruit	36	16	0.07	0.78	0.04	0.23	0.2	1.1	0.05	0.2	6	N	N	Tr
247	frozen	N	N	N	N	0.03	0.18	0.1	0.8	0.04	0.3	3	N	N	0
248	**Rice pudding**, canned	N	N	N	N	0.03	0.14	0.2	0.7	0.02	Tr	N	N	N	0
249	**Trifle**	70	33	0.17	0.40	0.06	0.13	0.3	0.9	0.06	0.4	8	0.34	3.1	4
250	*frozen*	(70)	(33)	(0.17)	(0.40)	Tr	0.07	0.1	0.5	0.02	0.2	4	(0.34)	(3.1)	(4)
251	*with Dream Topping*	44	18	0.16	0.35	0.06	0.13	0.3	0.9	0.06	0.4	8	0.34	3.1	4
252	*with fresh cream*	(70)	(33)	(0.17)	(0.40)	0.06	0.10	0.1	0.5	Tr	0.2	(8)	(0.34)	(3.1)	(4)

Butter and related fats

12-253 to 12-260

Composition of food per 100g

No. 12-	Food	Description and main data sources	Water g	Total nitrogen g	Protein g	Fat g	Carbo-hydrate g	Energy value kcal	kJ
253	**Ghee,** butter	5 assorted samples	0.1	Tr	Tr	99.8	Tr	898	3693
254	palm	5 samples of the same brand	0.1	Tr	Tr	99.7	Tr	897	3689
255	vegetable	5 samples, 2 different types	0.1	Tr	Tr	99.8	Tr	898	3693
256	**Butter**	Analysis and literature sources	15.6[a]	0.08	0.5	81.7[ab]	Tr	737	3031
257	**Margarine**	Mixed sample	16.0	0.03	0.2	81.6	1.0	739	3039
258	**Dairy/fat spread**	6 samples, Krona, Clover and Golden Churn	22.0	0.06	0.4	73.4	Tr	662	2723
259	**Low-fat spread**	4 samples, Gold, Delight, Outline and own brand	49.9	0.91	5.8	40.5	0.5	390	1605
260	**Very low fat spread**	i.e. Gold Lowest, manufacturer's data (St Ivel)	N	1.30	8.3	25.0	3.6	273	1128

[a] Unsalted butter contains 15.7g water and 82.7g fat per 100g

[b] 'Half-fat' butter spreads, e.g. Half-fat Anchor, Kerrygold Light, contain 39 - 40g fat per 100g

Butter and related fats

Carbohydrate fractions and fatty acids, g per 100g
Cholesterol, mg per 100g

No. 12-	Food	Starch	Total sugars	Individual sugars						Fatty acids			Cholest-erol
				Gluc	Fruct	Galact	Sucr	Malt	Lact	Satd	Mono unsatd	Poly unsatd	
253	**Ghee**, butter	0	Tr	0	0	0	0	0	Tr	66.0	24.1	3.4	280
254	palm	0	Tr	0	0	0	0	0	0	47.0	35.3	8.9	0
255	vegetable	0	Tr	0	0	0	0	0	0	N[a]	N[a]	N[a]	0
256	**Butter**	0	Tr	0	0	0	0	0	Tr	54.0	19.8	2.6	230
257	**Margarine**	0	1.0	0	0	0	0	0	1.0	N[b]	N[b]	N[b]	N[b]
258	**Dairy/fat spread**	0	Tr	0	0	0	0	0	Tr	28.1	29.9	11.3	105
259	**Low-fat spread**	0	0.5	0	0	0	0	0	0.5	11.2	17.6	9.9	6
260	**Very low fat spread**	Tr	3.6	0	0	0	0	0	3.6	6.5	N	N	N

[a] The fatty acid composition will depend on the type of oil

[b] Values depend on type of margarine. The average cholesterol content of polyunsaturated margarine is Tr, other soft margarines 163mg, and hard margarines 285mg per 100g

Butter and related fats

Inorganic constituents per 100g

No. 12-	Food	Na	K	Ca	Mg	P	Fe	Cu	Zn	S	Cl	Mn	Se	I
							mg						μg	
253	**Ghee**, butter	2	3	Tr	Tr	Tr	0.2	Tr	Tr	Tr	28	Tr	Tr	N
254	palm	1	1	Tr	Tr	Tr	0.1	0.21	Tr	Tr	N	Tr	N	Tr
255	vegetable	1	1	Tr	Tr	Tr	Tr	0.14	Tr	Tr	N	Tr	N	N
256	**Butter**	750[a]	15	15	2	24	0.2	0.03	0.1	9	1150[a]	Tr	Tr	38
257	**Margarine**	800	5	4	1	12	0.3	0.04	N	12	1200	Tr	Tr	26
258	**Dairy/fat spread**	760	43	14	2	18	Tr	Tr	Tr	N	1270	Tr	N	N
259	**Low-fat spread**	650	110	39	4	82	Tr	0.12	0.2	N	800	Tr	N	N
260	**Very low fat spread**	1050	630	N	N	N	N	N	N	N	N	N	N	N

[a] Unsalted butter contains 11mg Na and 17mg Cl per 100g

Butter and related fats

No. 12-	Food	Retinol µg	Carotene µg	Vitamin D µg	Vitamin E mg	Thiamin mg	Ribo-flavin mg	Niacin mg	Trypt 60 mg	Vitamin B6 mg	Vitamin B12 µg	Folate µg	Panto-thenate mg	Biotin µg	Vitamin C mg
253	**Ghee**, butter	675	500	1.90	3.31	0	Tr	Tr	Tr	Tr	Tr	0	Tr	Tr	0
254	palm	Tr	Tr	0	7.40	0	0	Tr	Tr	Tr	0	0	Tr	Tr	0
255	vegetable	Tr	Tr	0	10.27[a]	0	0	Tr	Tr	Tr	0	0	Tr	Tr	0
256	**Butter**	815	430	0.76	2.00	Tr	(0.02)	(0.01)	0.12	Tr	Tr	Tr	(0.04)	Tr	Tr
257	**Margarine**	780	750[b]	7.94	8.00	Tr	Tr	Tr	Tr	Tr	Tr	Tr	Tr	Tr	0
258	**Dairy/fat spread**	800	845	5.80	5.06	Tr	Tr	Tr	0.09	Tr	Tr	Tr	Tr	Tr	0
259	**Low-fat spread**	920	985	8.00	6.33	Tr	Tr	Tr	1.37	Tr	Tr	Tr	Tr	Tr	0
260	**Very low fat spread**	N	N	N	N	Tr	Tr	Tr	1.95	Tr	Tr	Tr	Tr	Tr	0

[a] The vitamin E content will vary according to type of oil

[b] Some brands may not contain β-carotene

Savoury dishes and sauces

12-261 to 12-274

Composition of food per 100g

No. 12-	Food	Description and main data sources	Water g	Total nitrogen g	Protein g	Fat g	Carbo-hydrate g	Energy value kcal	kJ
261	**Bread sauce**, *made with whole milk*	Recipe	76.3	0.69	4.2	5.1	12.6	110	463
262	*made with semi-skimmed milk*	Recipe	78.1	0.71	4.3	3.1	12.8	93	393
263	*made with skimmed milk*	Recipe	79.2	0.71	4.3	1.8	12.8	81	345
264	**Cauliflower cheese**	Recipe	78.6	0.95	5.9	6.9	5.1	105	438
265	**Cheese and potato pie**	Recipe. Ref. 6	73.4	0.77	4.8	8.1	12.6	139	584
266	**Cheese pastry**	Recipe. Ref. 6	14.4	2.15	13.2	34.1	37.4	500	2083
267	**Cheese pudding**	Recipe	68.4	1.61	10.1	10.9	8.5	170	709
268	**Cheese sauce**, *made with whole milk*	Recipe	66.9	1.26	8.0	14.6	9.0	197	819
269	*made with semi-skimmed milk*	Recipe	68.7	1.28	8.1	12.6	9.1	179	750
270	*made with skimmed milk*	Recipe	69.8	1.28	8.1	11.3	9.1	168	702
271	**Cheese sauce packet mix**	10 samples, 4 brands	3.9	3.08	19.3	19.7	40.9	408	1711
272	*made up with whole milk*	Recipe	77.2	0.84	5.3	6.1	9.3	110	462
273	*made up with semi-skimmed milk*	Recipe	79.2	0.86	5.4	3.8	9.5	90	383
274	*made up with skimmed milk*	Recipe	80.5	0.86	5.4	2.3	9.5	78	328

Savoury dishes and sauces

Carbohydrate fractions and fatty acids, g per 100g
Cholesterol, mg per 100g

No. 12-	Food	Starch	Total sugars	Individual sugars						Fatty acids			Cholest-erol
				Gluc	Fruct	Galact	Sucr	Malt	Lact	Satd	Mono unsatd	Poly unsatd	
261	**Bread sauce,** *made*												
	with whole milk	8.0	4.7	N	N	0	N	N	4.2	2.6	1.6	0.6	14
262	*made with semi-skimmed milk*	8.0	4.8	N	N	0	N	N	4.4	1.4	1.0	0.5	8
263	*made with skimmed milk*	8.0	4.8	N	N	0	N	N	4.4	0.5	0.6	0.5	4
264	**Cauliflower cheese**	2.1	3.0	1.0	0.8	0	0.1	Tr	1.2	3.4	2.1	1.0	16
265	**Cheese and potato pie**	11.0	1.6	0.2	0.1	0	0.3	0	1.0	3.6	2.8	1.3	48
266	**Cheese pastry**	36.5	0.9	0.3	0.3	0	0.1	0.1	Tr	15.3	12.5	4.4	65
267	**Cheese pudding**	5.3	3.1	N	N	0	N	N	2.8	5.9	3.5	0.6	115
268	**Cheese sauce,** *made*												
	with whole milk	4.6	4.3	Tr	Tr	0	Tr	Tr	4.3	7.5	4.7	1.7	39
269	*made with semi-skimmed milk*	4.6	4.5	Tr	Tr	0	Tr	Tr	4.4	6.3	4.1	1.6	33
270	*made with skimmed milk*	4.6	4.5	Tr	Tr	0	Tr	Tr	4.4	5.4	3.7	1.5	29
271	**Cheese sauce packet mix**	35.3	5.6	Tr	Tr	0	0.3	Tr	5.3	N	N	N	N
272	*made up with whole milk*	3.9	5.4	Tr	Tr	0	Tr	Tr	5.3	N	N	N	N
273	*made up with*												
	semi-skimmed milk	3.9	5.6	Tr	Tr	0	Tr	Tr	5.5	N	N	N	N
274	*made up with skimmed milk*	3.9	5.6	Tr	Tr	0	Tr	Tr	5.5	N	N	N	N

Savoury dishes and sauces

No. 12-	Food	Na	K	Ca	Mg	P	Fe	Cu	Zn	S	Cl	Mn	Se	I
						mg							µg	
261	**Bread sauce, made** *with whole milk*	480	140	120	16	96	0.3	0.03	0.4	40	760	0.1	6	15
262	*made with semi-skimmed milk*	480	150	120	16	99	0.3	0.03	0.5	41	760	0.1	6	15
263	*made with skimmed milk*	480	150	120	17	99	0.3	0.03	0.5	41	760	0.1	6	15
264	**Cauliflower cheese**	200	300	120	17	120	0.6	0.03	0.7	N	320	0.2	1	8
265	**Cheese and potato pie**	170	220	88	14	94	0.4	0.06	0.5	54	290	0.8	2	14
266	**Cheese pastry**	510	98	310	19	220	1.1	0.09	1.1	N	820	0.3	6	22
267	**Cheese pudding**	460	140	220	17	190	0.7	0.05	1.0	110	710	0.1	8	29
268	**Cheese sauce, made** *with whole milk*	450	150	240	17	180	0.2	0.02	0.8	N	710	Tr	3	23
269	*made with semi-skimmed milk*	450	150	240	17	180	0.2	0.02	0.8	N	710	Tr	3	23
270	*made with skimmed milk*	450	150	240	18	180	0.2	0.02	0.8	N	710	Tr	3	23
271	**Cheese sauce packet mix**	3680	330	430	36	730	0.8	Tr	2.5	N	4980	0.3	N	N
272	*made up with whole milk*	460	170	160	14	170	0.1	Tr	0.6	N	650	Tr	N	N
273	*made up with semi-skimmed milk*	460	190	170	14	170	0.1	Tr	0.7	N	650	Tr	N	N
274	*made up with skimmed milk*	460	190	170	15	170	0.1	Tr	0.7	N	650	Tr	N	N

No. 12-	Food	Retinol µg	Carotene µg	Vitamin D µg	Vitamin E mg	Thiamin mg	Ribo-flavin mg	Niacin mg	Trypt 60 mg	Vitamin B6 mg	Vitamin B12 µg	Folate µg	Panto-thenate mg	Biotin µg	Vitamin C mg
261	**Bread sauce**, *made with whole milk*	58	31	0.16	0.20	0.05	0.14	0.3	0.9	0.05	0.3	4	0.27	1.8	Tr
262	*made with semi-skimmed milk*	31	20	0.14	0.16	0.05	0.14	0.3	1.0	0.05	0.3	4	0.25	1.9	1
263	*made with skimmed milk*	14	12	0.14	0.14	0.05	0.14	0.3	1.0	0.05	0.3	4	0.25	1.9	1
264	**Cauliflower cheese**	63	79	0.22	0.41	0.10	0.10	0.5	1.4	0.16	0.2	7	0.40	1.8	7
265	**Cheese and potato pie**	86	55	0.52	0.52	0.09	0.09	0.3	1.2	0.11	0.4	12	0.28	2.2	3
266	**Cheese pastry**	205	165	1.04	1.28	0.12	0.13	0.8	3.0	0.08	0.4	10	0.20	1.5	Tr
267	**Cheese pudding**	130	50	0.46	0.40	0.05	0.24	0.3	2.5	0.07	0.9	11	0.54	6.3	Tr
268	**Cheese sauce**, *made with whole milk*	150	105	0.55	0.68	0.05	0.22	0.2	1.9	0.06	0.5	6	0.31	2.3	1
269	*made with semi-skimmed milk*	125	93	0.54	0.62	0.05	0.23	0.2	1.9	0.06	0.5	6	0.29	2.3	1
270	*made with skimmed milk*	105	85	0.53	0.60	0.05	0.23	0.2	1.9	0.06	0.5	6	0.29	2.3	1
271	**Cheese sauce packet mix**	140	150	0.54	1.84	0.12	0.58	0.9	4.1	0.26	1.3	46	N	N	0
272	*made up with whole milk*	67	37	0.09	0.29	0.04	0.23	0.2	1.2	0.07	0.5	5	N	N	1
273	*made up with semi-skimmed milk*	36	25	0.07	0.23	0.04	0.24	0.2	1.2	0.07	0.5	5	N	N	1
274	*made up with skimmed milk*	16	16	0.06	0.20	0.04	0.24	0.2	1.2	0.07	0.5	5	N	N	1

Savoury dishes and sauces continued

Composition of food per 100g

No. 12-	Food	Description and main data sources	Water g	Total nitrogen g	Protein g	Fat g	Carbo-hydrate g	Energy value kcal	kJ
275	**Macaroni cheese**	Recipe	67.1	1.18	7.3	10.8	13.6	178	743
276	canned	10 samples of the same brand (Heinz)	74.8	0.79	4.5	6.5	16.4	138	579
277	**Mayonnaise**	Recipe	16.9	0.31	1.9	79.3	0.2	724	2979
278	retail	6 samples, 5 brands	18.8	0.18	1.1	75.6	1.7	691	2843
279	retail, reduced calorie	6 samples, 5 brands	59.6	0.16	1.0	28.1	8.2	288	1188
280	**Onion sauce**, *made*								
	with whole milk	Recipe	80.4	0.45	2.8	6.5	8.3	99	414
281	*made with semi-skimmed milk*	Recipe	81.7	0.46	2.9	5.0	8.4	86	361
282	*made with skimmed milk*	Recipe	82.6	0.46	2.9	4.0	8.4	77	324
283	**Quiche**, cheese and egg	Recipe. Ref. 6	46.7	2.01	12.5	22.2	17.3	314	1310
284	cheese and egg, wholemeal	Recipe	46.7	2.12	13.2	22.4	14.5	308	1283
285	Lorraine	Recipe	33.8	2.59	16.1	28.1	19.8	391	1629
286	Lorraine, wholemeal	Recipe	33.8	2.71	16.9	28.3	16.6	384	1599
287	mushroom	Recipe	50.4	1.69	10.0	19.5	18.3	284	1185
288	mushroom, wholemeal	Recipe	50.4	1.81	10.8	19.7	15.2	277	1156
289	**Raita**, plain	Spiced curd/yogurt. Recipe. Ref. 5	74.3	0.42	2.6	15.3	5.5	166	690
290	yogurt and gram flour	Recipe. Ref. 5	78.1	0.56	3.5	6.5	8.9	106	441

Savoury dishes and sauces *continued*

Carbohydrate fractions and fatty acids, g per 100g
Cholesterol, mg per 100g

No. 12-	Food	Starch	Total sugars	Gluc	Fruct	Galact	Sucr	Malt	Lact	Satd	Mono-unsatd	Poly-unsatd	Cholest-erol
											Fatty acids		
				Individual sugars									
275	**Macaroni cheese**	10.7	2.9	Tr	Tr	0	0.1	0.1	2.7	5.6	3.4	1.2	28
276	canned	14.5	1.9	0.1	0.1	Tr	0.9	0.2	0.6	2.3	2.5	1.4	8
277	**Mayonnaise**	Tr	0.1	Tr	Tr	0	Tr	0	0	11.6	54.5	8.5	120
278	retail	0.4	1.3	0.1	0.1	0	1.1	0	0	11.1	17.3	43.9	75
279	retail, reduced calorie	3.6	4.6	1.1	1.0	0	2.5	0	0	N	N	N	N
280	**Onion sauce**, *made* *with whole milk*	3.5	4.7	0.7	0.4	0	0.3	Tr	3.3	2.8	2.2	1.1	16
281	*made with semi-skimmed milk*	3.5	4.9	0.7	0.4	0	0.3	Tr	3.4	1.8	1.8	1.1	11
282	*made with skimmed milk*	3.5	4.9	0.7	0.4	0	0.3	Tr	3.4	1.2	1.5	1.1	8
283	**Quiche**, cheese and egg	15.7	1.6	0.1	0.1	0	0.1	Tr	1.3	10.3	7.9	2.3	140
284	cheese and egg, wholemeal	12.7	1.7	0.1	0.1	0	0.2	0	1.3	10.4	8.0	2.4	140
285	Lorraine	17.5	2.3	0.1	0.1	0	0.1	Tr	2.0	12.3	10.9	3.2	140
286	Lorraine, wholemeal	14.2	2.4	0.1	0.1	0	0.2	0	2.0	12.3	10.9	3.3	140
287	mushroom	16.9	1.4	0.1	0.1	0	0.1	Tr	1.0	8.7	7.1	2.4	105
288	mushroom, wholemeal	13.7	1.5	0.1	0.1	0	0.2	0	1.0	8.7	7.1	2.5	105
289	**Raita**, plain	N	(2.6)	Tr	Tr	(1.0)	0	0	(1.6)	9.6	3.5	0.6	41
290	yogurt and gram flour	2.6	6.3	1.6	0.9	1.2	0.8	0	1.8	4.2	1.5	0.2	18

Savoury dishes and sauces *continued*

Inorganic constituents per 100g

No. 12-	Food	Na	K	Ca	Mg	P	Fe	Cu	Zn	S	Cl	Mn	Se	I
							mg						µg	
275	**Macaroni cheese**	310	110	170	17	150	0.4	0.05	0.8	N	490	0.1	4	16
276	canned	560	66	100	11	140	0.3	0.04	0.1	67	850	0.1	N	N
277	**Mayonnaise**	240	29	17	7	58	0.8	0.03	0.5	28	390	0.2	2	16
278	retail	450	16	8	1	27	0.3	0.02	0.1	N	750	Tr	N	35
279	retail, reduced calorie	(940)	N	N	N	N	N	N	N	N	(1450)	N	N	N
280	**Onion sauce**, *made* *with whole milk*	430	140	90	12	73	0.3	0.02	0.3	N	690	0.1	1	13
281	*made with semi-skimmed milk*	435	150	93	12	75	0.3	0.02	0.3	N	690	0.1	1	13
282	*made with skimmed milk*	430	150	93	12	75	0.3	0.02	0.4	N	690	0.1	1	13
283	**Quiche**, cheese and egg	340	120	260	17	220	1.0	0.06	1.1	N	530	0.1	7	31
284	cheese and egg, wholemeal	340	160	240	37	270	1.4	0.12	1.6	N	520	0.7	17	N
285	Lorraine	770	190	230	21	250	1.2	0.11	1.6	N	1260	0.1	N	30
286	Lorraine, wholemeal	770	240	200	43	300	1.6	0.18	2.1	N	1250	0.7	N	N
287	mushroom	290	160	200	15	180	1.0	0.19	0.9	N	470	0.2	7	24
288	mushroom, wholemeal	290	200	170	37	230	1.4	0.26	1.5	N	460	0.7	18	N
289	**Raita**, plain	180	190	96	16	78	2.2	N	N	N	N	N	N	N
290	yogurt and gram flour	450	220	95	19	88	1.0	0.05	0.5	47	710	Tr	N	27

Savoury dishes and sauces *continued*

No. 12-	Food	Retinol µg	Carotene µg	Vitamin D µg	Vitamin E mg	Thiamin mg	Ribo-flavin mg	Niacin mg	Trypt 60 mg	Vitamin B6 mg	Vitamin B12 µg	Folate µg	Panto-thenate mg	Biotin µg	Vitamin C mg
275	**Macaroni cheese**	110	74	0.36	0.45	0.04	0.16	0.3	1.7	0.05	0.4	5	0.26	1.6	Tr
276	canned	N	N	N	N	0.06	0.06	0.4	0.9	0.08	0.6	6	N	N	0
277	**Mayonnaise**	56	Tr	0.53	4.21	0.03	0.06	Tr	0.5	0.03	0.7	13	0.49	5.3	0
278	retail	86	100	0.33	18.94	0.02	0.07	Tr	0.3	0.01	0.5	4	N	N	N
279	retail, reduced calorie	N	N	N	N	N	N	N	0.3	N	N	N	N	N	N
280	**Onion sauce**, *made*														
	with whole milk	71	78	0.39	0.45	0.06	0.12	0.3	0.6	0.08	0.3	6	0.25	1.5	1
281	*made with semi-skimmed milk*	50	70	0.38	0.41	0.06	0.12	0.3	0.7	0.08	0.3	6	0.23	1.6	2
282	*made with skimmed milk*	36	64	0.37	0.39	0.06	0.12	0.3	0.7	0.08	0.3	6	0.23	1.6	2
283	**Quiche**, cheese and egg	185	100	0.93	0.91	0.08	0.23	0.4	3.1	0.08	1.0	13	0.53	6.6	Tr
284	cheese and egg, wholemeal	185	100	0.93	1.14	0.10	0.24	1.2	3.2	0.13	1.0	17	0.61	7.8	Tr
285	Lorraine	160	93	0.85	0.91	0.15	0.25	1.7	3.6	0.15	1.0	12	0.60	6.5	Tr
286	Lorraine, wholemeal	160	93	0.85	1.16	0.18	0.26	2.5	3.7	0.21	1.0	16	0.69	7.9	Tr
287	mushroom	150	87	0.82	0.83	0.09	0.22	1.0	2.5	0.09	0.7	14	0.67	7.0	Tr
288	mushroom, wholemeal	150	87	0.82	1.07	0.11	0.23	1.8	2.6	0.15	0.7	18	0.76	8.4	Tr
289	**Raita**, plain	145	975	0.14	0.35	0.04	0.12	0.4	0.5	0.03	0.1	4	0.15	1.0	1
290	yogurt and gram flour	43	210	0.12	0.20	0.08	0.10	0.4	0.7	0.08	0.1	9	0.27	1.4	3

Savoury dishes and sauces *continued*

Composition of food per 100g

No. 12-	Food	Description and main data sources	Water g	Total nitrogen g	Protein g	Fat g	Carbo-hydrate g	Energy value kcal	kJ
291	**Salad cream**	3 samples, different brands	47.2	0.23	1.5	31.0	16.7	348	1440
292	reduced calorie	Analysis and manufacturer's data	N	0.16	1.0	17.2	9.4	194	804
293	**Tzatziki**	Yogurt-based Greek starter. Recipe	85.8	0.59	3.7	4.9	2.0	66	275
294	**Welsh rarebit**	Recipe	31.6	2.24	13.8	26.3	23.8	380	1586
295	wholemeal	Recipe	31.3	2.31	14.3	25.6	22.1	369	1542
296	**White sauce**, savoury								
	made with whole milk	Recipe	73.7	0.66	4.1	10.3	10.9	150	624
297	*made with*								
	semi-skimmed milk	Recipe	75.8	0.68	4.2	7.8	11.1	128	539
298	*made with skimmed milk*	Recipe	77.2	0.68	4.2	6.2	11.1	114	480
299	*sweet, made with whole milk*	Recipe	68.3	0.61	3.8	9.5	18.6	170	712
300	*made with*								
	semi-skimmed milk	Recipe	70.2	0.62	3.9	7.2	18.8	150	634
301	*made with skimmed milk*	Recipe	71.5	0.62	3.9	5.7	18.8	138	580
302	**White sauce packet mix**	15 samples, 2 brands	4.3	1.59	9.0	10.5	60.0	355	1501
303	*made up with whole milk*	Recipe	80.5	0.62	3.9	4.7	9.4	93	389
304	*made up with*								
	semi-skimmed milk	Recipe	82.5	0.64	4.0	2.4	9.6	73	309
305	*made up with skimmed milk*	Recipe	83.8	0.64	4.0	0.9	9.6	59	254

Savoury dishes and sauces *continued*

Carbohydrate fractions and fatty acids, g per 100g
Cholesterol, mg per 100g

No. 12-	Food	Starch	Total sugars	Gluc	Fruct	Galact	Sucr	Malt	Lact	Satd	Mono-unsatd	Poly-unsatd	Cholesterol
						Individual sugars					Fatty acids		
291	**Salad cream**	Tr	16.7	1.9	1.9	0	12.9	0	0	3.9	6.2	19.4	43
292	reduced calorie	0.2	9.2	2.5	2.3	0	4.4	0	0	2.5	4.7	9.1	7
293	**Tzatziki**	0.3	1.7	0.3	0.4	0.6	0.1	0	0.3	2.9	1.4	0.2	N
294	**Welsh rarebit**	21.5	2.3	N	N	0	N	N	0.8	16.6	7.0	1.0	75
295	wholemeal	20.4	1.7	N	N	0	N	N	0.7	15.8	6.8	1.1	70
296	**White sauce**, savoury *made with whole milk*	5.6	5.3	Tr	Tr	0	Tr	Tr	5.2	4.4	3.5	1.8	26
297	*made with* *semi-skimmed milk*	5.6	5.5	Tr	Tr	0	Tr	Tr	5.4	2.9	2.8	1.7	18
298	*made with skimmed milk*	5.6	5.5	Tr	Tr	0	Tr	Tr	5.4	1.9	2.4	1.7	13
299	*sweet, made with whole milk*	5.2	13.5	Tr	Tr	0	8.6	Tr	4.8	4.1	3.3	1.7	24
300	*made with* *semi-skimmed milk*	5.2	13.7	Tr	Tr	0	8.6	Tr	5.0	2.7	2.6	1.6	17
301	*made with skimmed milk*	5.2	13.7	Tr	Tr	0	8.6	Tr	5.0	1.7	2.2	1.5	12
302	**White sauce packet mix**	53.8	6.2	Tr	Tr	0	2.5	0.2	3.5	N	N	N	N
303	*made up with whole milk*	4.1	5.3	Tr	Tr	0	0.2	Tr	5.1	N	N	N	N
304	*made up with* *semi-skimmed milk*	4.1	5.5	Tr	Tr	0	0.2	Tr	5.3	N	N	N	N
305	*made up with skimmed milk*	4.1	5.5	Tr	Tr	0	0.2	Tr	5.3	N	N	N	N

Savoury dishes and sauces *continued*

No. 12-	Food	Na	K	Ca	Mg	P	Fe	Cu	Zn	S	Cl	Mn	Se	I
							mg						μg	
291	**Salad cream**	1040	40	18	9	48	0.5	0.02	0.3	N	1620	0.1	N	11
292	reduced calorie	N	N	N	N	N	N	N	N	N	N	N	N	N
293	**Tzatziki**	370	150	88	12	91	0.3	0.01	0.3	N	570	N	1	N
294	**Welsh rarebit**	920	100	340	23	240	0.9	0.08	1.2	130	1420	0.3	18	26
295	wholemeal	910	170	300	52	290	1.5	0.15	1.8	130	1410	1.0	22	22
296	**White sauce**, savoury													
	made with whole milk	400	160	130	15	110	0.2	0.01	0.4	N	640	Tr	1	19
297	made with													
	semi-skimmed milk	400	170	140	15	110	0.2	0.01	0.5	N	640	Tr	1	19
298	made with skimmed milk	400	170	140	16	110	0.2	0.01	0.5	N	640	Tr	1	19
299	sweet, made with whole milk	110	150	120	12	98	0.2	0.01	0.4	N	190	Tr	1	17
300	made with													
	semi-skimmed milk	110	160	130	12	100	0.2	0.01	0.5	N	190	Tr	1	17
301	made with skimmed milk	110	160	130	13	100	0.2	0.01	0.5	N	190	Tr	1	17
302	**White sauce packet mix**	4160	310	98	28	170	0.6	Tr	0.8	N	7670	0.5	N	N
303	made up with whole milk	400	160	120	13	110	0.1	Tr	0.4	N	680	Tr	N	N
304	made up with													
	semi-skimmed milk	400	170	130	13	110	0.1	Tr	0.5	N	680	Tr	N	N
305	made up with skimmed milk	400	170	130	14	110	0.1	Tr	0.5	N	680	Tr	N	N

Savoury dishes and sauces *continued*

No. 12-	Food	Retinol µg	Carotene µg	Vitamin D µg	Vitamin E mg	Thiamin mg	Ribo-flavin mg	Niacin mg	Trypt 60 mg	Vitamin B6 mg	Vitamin B12 µg	Folate µg	Panto-thenate mg	Biotin µg	Vitamin C mg
291	**Salad cream**	9	17	0.19	10.47	N	N	N	0.3	0.03	0.5	3	N	N	0
292	reduced calorie	N	N	N	N	N	N	N	0.2	N	N	N	N	N	0
293	**Tzatziki**	60	(46)	0.03	0.20	0.03	0.20	0.1	0.9	0.04	0.1	7	N	N	1
294	**Welsh rarebit**	250	150	0.21	0.51	0.08	0.16	1.1	3.1	0.06	0.4	10	0.27	1.5	Tr
295	wholemeal	235	145	0.20	0.57	0.13	0.17	2.0	3.2	0.08	0.4	16	0.39	3.5	Tr
296	**White sauce**, savoury *made with whole milk*	115	77	0.62	0.71	0.05	0.18	0.2	0.9	0.06	0.4	4	0.32	2.1	Tr
297	*made with semi-skimmed milk*	79	64	0.60	0.64	0.05	0.19	0.2	1.0	0.06	0.4	4	0.29	2.2	1
298	*made with skimmed milk*	58	55	0.59	0.61	0.05	0.19	0.2	1.0	0.06	0.4	4	0.29	2.2	1
299	*sweet, made with whole milk*	105	71	0.57	0.65	0.05	0.17	0.2	0.9	0.06	0.4	3	0.29	1.9	Tr
300	*made with semi-skimmed milk*	73	59	0.55	0.59	0.05	0.18	0.2	0.9	0.06	0.4	3	0.27	2.0	1
301	*made with skimmed milk*	54	51	0.54	0.56	0.05	0.18	0.2	0.9	0.06	0.4	3	0.27	2.0	1
302	**White sauce packet mix**	Tr	Tr	0	2.30	0.21	0.30	1.8	1.9	0.23	Tr	3	N	N	0
303	*made up with whole milk*	52	21	0.03	0.26	0.04	0.19	0.2	0.9	0.06	0.4	3	N	N	1
304	*made up with semi-skimmed milk*	21	9	0.01	0.20	0.04	0.20	0.2	0.9	0.06	0.4	3	N	N	1
305	*made up with skimmed milk*	1	Tr	Tr	0.17	0.04	0.20	0.2	0.9	0.06	0.4	3	N	N	1

Eggs

and

Egg Dishes

Eggs

Composition of food per 100g

No. 12-	Food	Description and main data sources	Water g	Total nitrogen g	Protein g	Fat g	Carbo-hydrate g	Energy value kcal	kJ
801	**Eggs**, chicken, whole, raw[a]	Analysis of battery, deep litter and free range	75.1	2.01	12.5	10.8	Tr	147	612
802	battery, raw	Mixed sample	74.9	2.03	12.7	10.7	Tr	147	612
803	free-range, raw	As identified by free range egg scheme	75.2	1.99	12.4	10.9	Tr	148	614
804	chicken, white, raw	34 eggs and literature sources	88.3	1.44	9.0	Tr	Tr	36	153
805	chicken, yolk, raw	34 eggs and literature sources	51.0	2.58	16.1	30.5	Tr	339	1402
806	chicken, boiled	10 eggs	75.1	2.01	12.5	10.8	Tr	147	612
807	dried	Calculated from whole raw egg	4.0	7.75	48.4	41.6	Tr	568	2362
808	fried, with fat	12 eggs, shallow fried in vegetable oil	70.1	2.18	13.6	13.9	Tr	179	745
809	fried, without fat	12 eggs fried in non-stick pan	70.6	2.40	15.0	12.7	Tr	174	725
810	poached[b]	10 eggs, no fat added	75.1	2.01	12.5	10.8	Tr	147	612
811	scrambled, with milk	Recipe	62.4	1.71	10.7	22.6	0.6	247	1025
812	scrambled, without milk	12 eggs scrambled in non-stick pan	73.2	2.21	13.8	11.6	Tr	160	664
813	duck, whole, raw	Analytical and literature sources. Ref. 3	70.6	2.29	14.3	11.8	Tr	163	680
814	duck, boiled and salted	20 eggs from 7 different Chinese shops	62.0	2.34	14.6	15.5	Tr	198	822
815	quail, whole, raw	Analytical and literature sources. Ref. 3	72.4	2.06	12.9	11.1	Tr	151	630
816	turkey, whole, raw	Literature sources. Ref. 3	72.5	2.20	13.7	12.2	Tr	165	684

[a] An average egg is composed of 11% shell, 58% white and 31% yolk

[b] Eggs poached with fat added contain 74.4g water, 12.4g protein, 11.7g fat, Tr carbohydrate, 155 kcals and 644 kJ per 100g

Carbohydrate fractions and fatty acids, g per 100g
Cholesterol, mg per 100g

| No. 12- | Food | Starch | Total sugars | Individual sugars | | | | | | Fatty acids | | | Cholest-erol |
				Gluc	Fruct	Galact	Sucr	Malt	Lact	Satd	Mono-unsatd	Poly-unsatd	
801	**Eggs**, chicken, whole, raw	0	Tr	Tr	0	0	0	0	0	3.1	4.7	1.2	385
802	battery, raw	0	Tr	Tr	0	0	0	0	0	3.1	4.6	1.2	380
803	free-range, raw	0	Tr	Tr	0	0	0	0	0	2.9	4.9	1.2	390
804	chicken, white, raw	0	Tr	Tr	0	0	0	0	0	Tr	Tr	Tr	0
805	chicken, yolk, raw	0	Tr	Tr	0	0	0	0	0	8.7	13.2	3.4	1120
806	chicken, boiled	0	Tr	Tr	0	0	0	0	0	3.1	4.7	1.2	385
807	dried	0	Tr	Tr	0	0	0	0	0	11.9	18.0	4.6	1500
808	fried, with fat	0	Tr	Tr	0	0	0	0	0	N[a]	N[a]	N[a]	N[a]
809	fried, without fat	0	Tr	Tr	0	0	0	0	0	3.6	5.5	1.4	450
810	poached	0	Tr	Tr	0	0	0	0	0	3.1	4.7	1.2	385
811	scrambled, with milk	0	0.6	Tr	0	0	0	0	0.6	11.6	7.2	1.4	350
812	scrambled, without milk	0	Tr	Tr	0	0	0	0	0	3.3	5.0	1.3	415
813	duck, whole, raw	0	Tr	Tr	0	0	0	0	0	2.9	4.9	2.0	680
814	duck, boiled and salted	0	Tr	Tr	0	0	0	0	0	3.8	6.4	2.7	890
815	quail, whole, raw	0	Tr	Tr	0	0	0	0	0	3.1	4.9	1.3	900
816	turkey, whole, raw	0	Tr	Tr	0	0	0	0	0	3.7	4.7	1.4	680

[a] Dependent on the type of fat used

No. 12-	Food	mg											µg	
		Na	K	Ca	Mg	P	Fe	Cu	Zn	S	Cl	Mn	Se	I
801	**Eggs**, chicken, whole, *raw*	140	130	57	12	200	1.9	0.08	1.3	180	160	Tr	11	53
802	battery, *raw*	140	130	59	12	200	2.0	0.08	1.3	180	160	Tr	11	53
803	free-range, *raw*	140	130	56	11	200	1.7	0.08	1.3	180	160	Tr	11	53
804	chicken, white, *raw*	190	150	5	11	33	0.1	0.02	0.1	180	170	Tr	6	(3)
805	chicken, yolk, *raw*	50	120	130	15	500	6.1	0.15	3.9	170	140	0.1	20	(140)
806	chicken, *boiled*	140	130	57	12	200	1.9	0.08	1.3	180	160	Tr	11	53
807	*dried*	540	500	220	46	770	7.3	0.31	5.0	690	620	0.1	42	200
808	*fried, with fat*	160	150	65	14	230	2.2	0.09	1.5	200	180	Tr	12	60
809	*fried, without fat*	170	150	67	14	240	2.2	0.09	1.5	210	190	Tr	13	63
810	*poached*	140	130	57	12	200	1.9	0.08	1.3	180	160	Tr	11	53
811	*scrambled, with milk*	1030	130	63	17	180	1.6	0.07	1.1	150	1550	Tr	9	52
812	*scrambled, without milk*	150	140	61	13	210	2.0	0.09	1.4	190	170	Tr	12	57
813	duck, whole, *raw*	120	190	63	16	200	2.9	(0.50)	1.4	N	N	(0.1)	N	N
814	duck, *boiled and salted*	1690	800	99	13	270	3.2	0.53	3.5	N	2920	0.1	N	N
815	quail, whole, *raw*	N	N	64	N	230	3.7	N	N	N	N	N	N	N
816	turkey, whole, *raw*	N	N	99	N	170	4.1	N	N	N	N	N	N	N

Eggs

No. 12-	Food	Retinol µg	Carotene µg	Vitamin D µg	Vitamin E mg	Thiamin mg	Ribo-flavin mg	Niacin mg	Trypt 60 mg	Vitamin B6 mg	Vitamin B12 µg	Folate µg	Panto-thenate mg	Biotin µg	Vitamin C mg
801	**Eggs**, chicken, whole, raw	190	Tr	1.75[a]	1.11	0.09	0.47	0.07	3.69	0.12	2.5	50	1.77	20.0	0
802	battery, raw	190	Tr	1.75	1.11	0.09	0.47	0.07	3.72	0.12	2.4	51	1.74	20.0	0
803	free-range, raw	190	Tr	1.75	1.11	0.09	0.45	0.07	3.65	0.12	2.7	49	1.80	20.0	0
804	chicken, white, raw	0	0	0	0	0.01	0.43	0.09	2.64	0.02	0.1	13	0.30	7.0	0
805	chicken, yolk, raw	535	Tr	4.94	3.11	0.30	0.54	0.06	4.73	0.30	6.9	130	4.60	50.0	0
806	chicken, boiled	190	Tr	1.75	1.11	0.07	0.35	0.07	3.69	0.12	1.1	39	1.30	16.0	0
807	dried	730	Tr	6.75	4.28	0.31	1.81	0.27	14.21	0.46	7.2	190	6.90	69.4	0
808	fried, with fat	215	Tr	1.99	N	0.07	0.31	0.08	4.00	0.14	1.6	(40)	1.30	18.0	0
809	fried, without fat	225	Tr	2.07	1.31	0.08	0.36	0.09	4.40	0.14	1.2	48	1.50	21.0	0
810	poached	190	Tr	1.75	1.11	0.07	0.36	0.07	3.69	0.12	1.0	45	1.30	15.0	0
811	scrambled, with milk	295	72	1.55	1.23	0.07	0.33	0.07	3.12	0.09	2.1	28	1.29	16.5	Tr
812	scrambled, without milk	205	Tr	1.88	1.19	0.07	0.47	0.08	4.05	0.13	1.2	44	1.10	16.0	0
813	duck, whole, raw	540	120	5.00	N	0.16	0.47	0.17	4.20	0.25	5.4	80	N	N	0
814	duck, boiled and salted	85	N	N	N	0.16	0.52	0.10	4.29	N	3.5	28	N	N	0
815	quail, whole, raw	N	N	N	N	0.13	0.79	0.15	3.78	0.15	N	N	N	N	0
816	turkey, whole, raw	N	N	N	N	0.11	0.47	0.02	4.03	N	N	N	N	N	0

a If the hens have been fed a supplement, values may be considerably higher

Egg dishes

Composition of food per 100g

No. 12-	Food	Description and main data sources	Water g	Total nitrogen g	Protein g	Fat g	Carbo-hydrate g	Energy value kcal	kJ
Savoury egg dishes									
817	**Curried omelette/egg** masala	Recipe. Ref. 5	46.0	1.35	8.4	36.6	4.1	377	1560
818	**Egg fried rice**	Recipe	60.7	0.69	4.2	10.6	25.7	208	873
819	**Egg fu yung**	Recipe	63.3	1.68	9.9	20.6	2.2	239	991
820	**Egg nog**	Recipe	81.2	0.60	3.8	4.1	8.8	105	439
821	**Omelette**, plain	Recipe	69.0	1.75	10.9	16.4	Tr	191	792
822	cheese	Recipe. Ref. 6	57.7	2.52	15.9	22.6	Tr	266	1106
823	Spanish	Recipe	77.1	0.91	5.7	8.3	6.2	120	501
824	**Scotch eggs**, retail	10 samples, 8 brands	54.0	1.92	12.0	17.1	13.1	251	1046
825	homemade	Recipe	54.0	1.88	11.6	20.9	11.7	278	1157
826	**Souffle**, plain	Recipe. Ref. 6	66.5	1.23	7.6	14.7	10.4	201	838
827	cheese	Recipe	58.3	1.82	11.4	19.2	9.3	253	1053
Sweet egg dishes									
828	**Macaroon**	Recipe	4.2	1.58	8.6	19.6	62.4	444	1869
829	**Meringue**	Recipe	2.2	0.85	5.3	Tr	95.4	379	1616
830	*with cream*	Recipe. Ref. 6	34.1	0.53	3.3	23.6	40.0	376	1570

Carbohydrate fractions and fatty acids, g per 100g
Cholesterol, mg per 100g

No. 12-	Food	Starch	Total sugars	Individual sugars						Fatty acids			Cholest- erol
				Gluc	Fruct	Galact	Sucr	Malt	Lact	Satd	Mono- unsatd	Poly- unsatd	
Savoury egg dishes													
817	**Curried omelette/egg masala**	N	N	N	N	N	N	N	N	21.3	9.9	1.7	305
818	**Egg fried rice**	24.8	0.9	0.4	0.2	0	0.2	0	0	1.5	4.1	4.2	70
819	**Egg fu yung**	0.4	1.8	0.7	0.5	0	0.6	Tr	0	3.2	9.8	5.8	225
820	**Egg nog**	0	8.8	0.1	0.1	0	5.1	0	3.4	2.1	1.4	0.2	55
821	**Omelette**, plain	0	Tr	Tr	0	0	0	0	0	7.4	5.8	1.3	355
822	cheese	0	Tr	Tr	0	0	0	0	Tr	12.2	7.2	1.2	265
823	Spanish	2.6	3.7	1.6	1.2	0	0.9	0	0	1.6	3.3	2.5	130
824	**Scotch eggs**, retail	13.1	Tr	Tr	0	0	0	0	0	4.3	6.6	3.3	165
825	homemade	11.5	0.2	Tr	Tr	0	Tr	Tr	0	7.7	9.6	1.8	190
826	**Souffle**, plain	7.7	2.7	0.1	0.1	0	Tr	Tr	2.5	5.0	5.7	2.8	175
827	cheese	6.8	2.4	0.1	0.1	0	Tr	Tr	2.3	8.3	6.8	2.7	175
Sweet egg dishes													
828	**Macaroon**	3.2	59.0	Tr	0	0	59.2	Tr	0	1.5	13.4	3.7	0
829	**Meringue**	0	95.4	Tr	0	0	95.4	0	0	Tr	Tr	Tr	0
830	*with cream*	0	40.0	Tr	0	0	38.2	0	1.9	14.7	6.8	0.1	65

No. 12-	Food	Na	K	Ca	Mg	P	mg Fe	Cu	Zn	S	Cl	Mn	µg Se	I
	Savoury egg dishes													
817	**Curried omelette/egg masala**	920	200	71	27	150	2.9	0.07	1.1	120	1380	Tr	N	N
818	**Egg fried rice**	27	72	13	5	67	0.5	0.07	0.7	60	38	0.3	5	10
819	**Egg fu yung**	85	220	65	37	180	2.0	0.15	1.2	N	110	0.2	8	33
820	**Egg nog**	56	120	89	9	90	0.3	0.01	0.4	43	90	Tr	2	17
821	**Omelette**, plain	1030	110	51	16	170	1.7	0.07	1.1	160	1530	Tr	9	50
822	cheese	900	100	280	19	280	1.2	0.06	1.5	180	1360	Tr	10	46
823	Spanish	120	210	31	13	97	1.1	0.06	0.6	N	190	0.2	4	20
824	**Scotch eggs**, retail	670	130	50	15	170	1.8	0.23	1.2	N	980	0.2	N	17
825	homemade	480	150	57	13	180	1.6	0.20	1.2	140	640	N	N	25
826	**Souffle**, plain	240	140	95	13	140	1.0	0.05	0.8	N	360	0.1	5	33
827	cheese	440	140	210	17	210	1.0	0.05	1.1	N	670	0.1	7	36
	Sweet egg dishes													
828	**Macaroon**	47	350	93	98	170	1.6	0.07	1.3	100	42	0.7	3	5
829	**Meringue**	110	90	4	6	19	0.1	0.03	0.2	110	100	Tr	3	2
830	*with cream*	70	84	39	6	42	Tr	0.01	0.3	55	75	Tr	1	1

No. 12-	Food	Retinol μg	Carotene μg	Vitamin D μg	Vitamin E mg	Thiamin mg	Ribo-flavin mg	Niacin mg	Trypt 60 mg	Vitamin B6 mg	Vitamin B12 μg	Folate μg	Panto-thenate mg	Biotin μg	Vitamin C mg
Savoury egg dishes															
817	**Curried omelette/egg masala**	405	205	1.29	1.36	0.09	0.24	0.4	2.3	0.10	1.4	22	0.92	11.8	2
818	**Egg fried rice**	34	8	0.31	0.20	0.03	0.08	0.3	1.1	0.06	0.4	8	0.39	4.5	Tr
819	**Egg fu yung**	110	18	1.03	2.61	0.09	0.31	0.5	2.6	0.10	1.5	36	1.13	N	1
820	**Egg nog**	59	14	0.23	0.18	0.04	0.17	0.1	1.0	0.05	0.6	9	0.44	3.8	Tr
821	**Omelette**, plain	235	37	1.58	1.13	0.07	0.33	0.1	3.2	0.09	2.2	30	1.33	17.3	0
822	cheese	265	100	1.13	0.92	0.06	0.32	0.1	4.2	0.09	1.8	27	0.98	12.4	Tr
823	Spanish	65	1615	0.60	0.88	0.09	0.14	0.5	1.5	0.14	0.9	22	0.64	7.1	21
824	**Scotch eggs**, retail	30	Tr	0.73	N	0.08	0.21	1.0	2.9	0.13	0.5	42	(1.10)	(8.7)	N
825	homemade	81	Tr	0.75	0.48	0.06	0.23	1.5	2.9	0.07	1.3	16	0.83	10.7	0
826	**Souffle**, plain	180	86	1.52	1.22	0.07	0.22	0.2	2.1	0.08	1.1	17	0.75	9.1	Tr
827	cheese	220	140	1.39	1.27	0.07	0.26	0.2	2.9	0.08	1.2	19	0.72	8.6	Tr
Sweet egg dishes															
828	**Macaroon**	0	0	0	7.32	0.07	0.37	0.8	1.7	0.04	Tr	19	0.20	1.8	0
829	**Meringue**	0	0	0	0	Tr	0.24	0.1	1.6	0.01	0.1	6	0.16	4.0	0
830	*with cream*	340	160	0.13	0.54	0.01	0.20	Tr	0.9	0.03	0.1	6	0.20	2.4	Tr

FIBRE FRACTIONS

Milk and eggs contain no fibre, but a number of dairy and egg products contain cereals. The weight of each of the main fibre fractions in these products has been calculated for the table below. Non-starch polysaccharides determined by the Englyst method (Englyst and Cummings, 1988) include cellulose and insoluble polysaccharides (insoluble fibre) together with soluble non-cellulosic polysaccharides (soluble fibre). Fibre determined by the Southgate method (Wenlock *et al.*, 1985) also includes lignins and some starch.

		Fibre and fibre fractions, g per 100g					
		Dietary fibre			Fibre fractions		
					Non-cellulosic polysaccharide		
		Southgate method	Englyst method	Cellulose	Soluble	Insoluble	Lignin
Ice creams							
199	**Banana split**	1.4	0.6	0.1	0.3	0.1	N
213	**Knickerbocker glory**	0.5	0.2	N	N	N	N
214	**Kulfi**	1.0	0.6	0.1	0.1	0.3	N
215	**Peach melba**	1.8	0.4	0.1	0.2	0.1	N
Puddings and chilled desserts							
218	**Cheesecake**	0.7	0.4	0.1	0.2	0.1	Tr
226	**Custard**, confectioners'	0.3	0.2	Tr	0.1	0.1	Tr
232	**Fruit fool**	2.2	1.2	0.4	0.4	0.4	N
233	**Instant dessert powder**	1.0	N	0.1	N	N	0.1
234	*made up with whole milk*	0.2	N	Tr	N	N	Tr
235	*made up with semi-skimmed milk*	0.2	N	Tr	N	N	Tr
236	*made up with skimmed milk*	0.2	N	Tr	N	N	Tr
241	**Milk pudding**, *made with whole milk*	0.2	0.1	N	N	N	N
235	*made with semi-skimmed milk*	0.2	0.1	N	N	N	N
236	*made with skimmed milk*	0.2	0.1	N	N	N	N
249	**Trifle**	0.5	0.5	0.1	0.1	0.2	N
251	*with Dream Topping*	0.5	0.5	0.1	0.1	0.2	N
Savoury dishes and sauces							
261	**Bread sauce**, *made with whole milk*	0.6	0.3	Tr	0.2	0.1	0.1
262	*made with semi-skimmed milk*	0.6	0.3	Tr	0.2	0.1	0.1
263	*made with skimmed milk*	0.6	0.3	Tr	0.2	0.1	0.1
264	**Cauliflower cheese**	1.4	1.3	0.3	0.6	0.4	N
265	**Cheese and potato pie**	0.6	0.7	0.3	0.4	0.1	Tr

Fibre and fibre fractions, g per 100g

		Dietary fibre		Fibre fractions			
					Non-cellulosic polysaccharide		
		Southgate method	Englyst method	Cellulose	Soluble	Insoluble	Lignin
266	**Cheese pastry**	1.7	1.5	0.1	0.7	0.7	Tr
267	**Cheese pudding**	0.4	0.2	Tr	0.1	0.1	0.1
268	**Cheese sauce**, *made with whole milk*	0.2	0.2	Tr	0.1	0.1	Tr
269	*made with semi-skimmed milk*	0.2	0.2	Tr	0.1	0.1	Tr
270	*made with skimmed milk*	0.2	0.2	Tr	0.1	0.1	Tr
275	**Macaroni cheese**	0.8	0.5	Tr	0.3	0.2	Tr
280	**Onion sauce**, *made with whole milk*	0.6	0.4	0.1	0.2	0.1	Tr
281	*made with semi-skimmed milk*	0.6	0.4	0.1	0.2	0.1	Tr
282	*made with skimmed milk*	0.6	0.4	0.1	0.2	0.1	Tr
283	**Quiche**, cheese and egg	0.7	0.6	Tr	0.3	0.3	Tr
284	wholemeal	1.8	1.9	0.3	0.4	1.2	0.1
285	Lorraine	0.8	0.7	Tr	0.3	0.3	Tr
286	wholemeal	2.0	2.1	0.3	0.5	1.3	0.1
287	mushroom	1.2	0.9	0.1	0.4	0.5	Tr
288	wholemeal	2.3	2.2	0.3	0.5	1.4	0.1
290	**Raita**, yogurt and gram flour	1.2	1.0	0.3	0.5	0.2	N
293	**Tzatziki**	0.2	0.2	0.1	0.1	Tr	Tr
294	**Welsh rarebit**	1.8	0.7	Tr	0.4	0.3	0.3
295	wholemeal	3.8	2.6	0.5	0.7	1.4	0.3
296	**White sauce**, savoury						
	made with whole milk	0.3	0.2	Tr	0.1	0.1	Tr
297	*made with semi-skimmed milk*	0.3	0.2	Tr	0.1	0.1	Tr
298	*made with skimmed milk*	0.3	0.2	Tr	0.1	0.1	Tr
299	sweet, *made with whole milk*	0.2	0.2	Tr	0.1	0.1	Tr
300	*made with semi-skimmed milk*	0.2	0.2	Tr	0.1	0.1	Tr
301	*made with skimmed milk*	0.2	0.2	Tr	0.1	0.1	Tr
Egg dishes							
818	**Egg fried rice**	0.9	0.4	0.2	0.1	0.1	Tr
819	**Egg fu yung**	1.6	1.3	0.4	0.2	0.6	Tr
823	**Omelette**, Spanish	2.0	1.4	0.7	0.5	0.2	N
825	**Scotch eggs**, homemade	0.4	0.2	Tr	0.1	0.1	Tr
826	**Souffle**, plain	0.4	0.3	Tr	0.2	0.2	Tr
827	cheese	0.3	0.3	Tr	0.1	0.1	Tr
828	**Macaroon**	4.8	2.7	0.7	0.4	1.6	Tr

RETINOL FRACTIONS

The main forms of retinol in dairy products and eggs are all-*trans* retinol and 13-*cis* retinol. The latter has about 75 per cent of the biological activity of the former. Eggs also contain 21 µg retinaldehyde per 100g; retinaldehyde has about 90 per cent of the activity of all-*trans* retinol.

	all-*trans*	13-*cis*			all-*trans*	13-*cis*
	µg per 100g				µg per 100g	

Cows milk

		all-*trans*	13-*cis*
3	**Skimmed milk**, pasteurised		
	fortified *plus SMP*	40	4
5	UHT, *fortified*	49	15
6	fortified *plus SMP*	39	15
10	**Semi-skimmed milk**, pasteurised		
	fortified *plus SMP*	84	8
13	**Whole milk,** pasteurised	48	4
14	*summer*	58	4
15	*winter*	38	4
16	UHT	44	5
18	**Channel Island milk**,		
	whole, pasteurised	43	3
19	*summer*	60	6
20	*winter*	25	1
24	**Calcium-fortified milk**,		
	Calcia	6	Tr
25	Vital	Tr	Tr
28	**Condensed milk**,		
	skimmed, *sweetened*	23	7
29	whole, *sweetened*	100	15
30	**Dried skimmed milk**	317	43
31	*with vegetable fat*	373	30
33	**Evaporated milk**, whole	93	20

Other milks

		all-*trans*	13-*cis*
37	**Goats milk**, pasteurised	40	6
41	**Sheeps milk**, *raw*	81	2

Infant formulas *(as powder)*

		all-*trans*	13-*cis*
44	**Aptamil**	805	80
46	**Cow & Gate Premium**	797	95
48	**Farley's Oster Milk**	920	110
52	**Cow & Gate Plus**	673	75
54	**Farley's Oster Milk Two**	683	125
56	**Milumil**	705	83

Milk-based drinks

		all-*trans*	13-*cis*
103	**Milk shake**, *purchased*	19	2

Creams

		all-*trans*	13-*cis*
112	**Half**	173	23
113	**Single**	295	26
115	**Whipping**	552	18
116	**Double**	586	18
117	**Clotted**	655	66
120	**Sterilised**, canned	188	69
122	**Single**	225	39
124	**Canned spray**	331	52

Cheeses

		all-*trans*	13-*cis*
131	**Brie**	216	91
132	**Caerphilly**	247	89
133	**Camembert**	193	48
135	**Cheddar**, Australian	216	68
136	Canadian	203	77
137	English	300	56
138	Irish	232	157
139	New Zealand	218	81
140	vegetarian	304	110
141	**Cheddar-type**, *reduced fat*	132	43
142	**Cheese spread**, plain	239	46
145	**Cheshire**	251	130
146	**Cheshire-type**, *reduced fat*	113	47
147	**Cottage cheese**, plain	34	13
151	**Danish blue**	238	59
152	**Derby**	273	92
153	**Double Gloucester**	284	78
154	**Edam**	132	58
157	**Feta**	170	64
158	**Fromage frais**, plain	96	6
162	**Goats milk soft cheese**	233	104
163	**Gouda**	195	70

Retinol fractions *continued*

		all-trans	13-cis
		µg per 100g	
166	**Lancashire**	263	80
167	**Leicester**	277	55
168	**Lymeswold**	368	95
170	**Mozzarella**	232	67
171	**Parmesan**	267	102
172	**Processed cheese**, *plain*	220	67
176	**Ricotta**	146	49
177	**Roquefort**	252	54
178	**Sage Derby**	281	83
181	**Stilton**, white	250	84
182	**Wensleydale**	224	70

Yogurts

184	**Whole milk yogurt**, plain	25	5
185	fruit	35	6
191	**Low fat yogurt**, muesli/nut	(7)	(1)
194	**Greek yogurt**, cows	101	18
195	sheep	71	20
196	**Soya yogurt**	18.	6

Ice creams

203	**Frozen ice cream desserts**	2	Tr
204	**Ice cream**, dairy, vanilla	100	19
206	non-dairy, vanilla	1	Tr
208	mixes	1	1

Puddings and chilled desserts

		all-trans	13-cis
		µg per 100g	
220	**Creme caramel**	32	7
244	**Mousse**, chocolate	38	11
246	fruit	33	4

Butter and related fats

253	**Ghee**, butter	502	229
256	**Butter**	761	71
258	**Dairy/fat spread**	773	33
259	**Low-fat spread**	894	32

Savoury dishes and sauces

271	**Cheese sauce packet mix**	9	5
278	**Mayonnaise**, retail	72	19
291	**Salad cream**	6	4

Eggs

801	**Eggs**, chicken, whole, *raw*	132	53

Egg dishes

824	**Scotch eggs**, retail	22	10

VITAMINS IN INFANT FORMULAS

Manufacturers may add more of some vitamins than is stated on their labels to ensure that the declared amount will always be present even if there are losses on storage. Where possible, the main tables give the values found during recent analyses of the products. The table below is based on the amounts stated by manufacturers in May 1989, converted for the reconstituted formulas to amounts per 100g (specific gravity = 1.03). The formulations may change in the future.

Vitamins per 100g

No. 12-	Food	Retinol μg	Carotene μg	Vitamin D μg	Vitamin E mg	Vitamin K μg	Thiamin mg	Riboflavin mg	Niacin mg	Vitamin B6 mg	Vitamin B12 μg	Folate μg	Pantothenate mg	Biotin μg	Vitamin C mg
44	Aptamil	465	Tr	7.80	5.0	31	0.31	0.39	3.1	0.23	1.2	78	3.10	8.7	46
45	Aptamil reconstituted	59	Tr	0.97	0.68	3.9	0.04	0.05	0.4	0.03	0.1	10	0.39	1.1	6
46	Cow & Gate Premium	600	Tr	8.70	9.00	40	0.30	0.80	3.0	0.30	1.6	80	2.0	12.0	60
47	Cow & Gate Premium reconstituted	78	Tr	1.07	1.07	4.9	0.04	0.10	0.4	0.04	0.2	9	0.25	1.5	7
48	Farley's Oster Milk	800	Tr	8.00	3.70	21	0.36	0.42	5.3	0.27	1.1	26	1.80	7.9	53
49	Farley's Oster Milk reconstituted	97	Tr	0.97	0.47	2.6	0.04	0.05	0.7	0.03	0.1	3	0.22	1.0	7
50	Gold Cap SMA	865	94	10.10	9.30	63	1.07	1.91	4.5	0.56	1.0	57	2.50	17.3	85
51	Gold Cap SMA reconstituted	105	12	1.25	1.15	7.9	0.13	0.24	0.6	0.07	0.1	7	0.31	2.1	11
52	Cow & Gate Plus	600	Tr	8.40	8.10	40	0.30	0.80	3.0	0.30	1.5	80	2.0	12.0	60
53	Cow & Gate Plus reconstituted	78	Tr	1.07	1.07	4.9	0.04	0.10	0.4	0.04	0.2	10	0.25	1.5	7
54	Farley's Oster Milk Two	700	Tr	7.20	3.30	19	0.31	0.38	4.7	0.24	0.9	23	1.6	7.0	46
55	Farley's Oster Milk Two reconstituted	94	Tr	0.97	0.45	2.5	0.04	0.05	0.6	0.03	0.1	3	0.21	0.9	6
56	Milumil	405	Tr	7.30	5.70	29	0.23	0.35	1.7	0.30	1.5	36	1.7	8.1	54
57	Milumil reconstituted	55	Tr	0.97	0.78	4.0	0.03	0.05	0.2	0.04	0.2	5	0.23	1.1	7

Vitamins in Infant Formulas *continued*

Vitamins per 100g

No. 12-	Food	Retinol µg	Carotene µg	Vitamin D µg	Vitamin E mg	Vitamin K µg	Thiamin mg	Riboflavin mg	Niacin mg	Vitamin B6 mg	Vitamin B12 µg	Folate µg	Pantothenate mg	Biotin µg	Vitamin C mg
58	**White Cap SMA**	855	93	10.00	9.00	64	1.04	1.48	4.4	0.50	1.0	58	2.52	18.0	85
59	**White Cap SMA** reconstituted	105	11	1.24	1.12	7.9	0.13	0.18	0.5	0.06	0.1	7	0.31	2.2	11
60	**Farley's Oster Soy**	730	Tr	8.00	3.70	20	0.31	0.41	5.1	0.26	1.1	26	1.80	8.0	51
61	**Farley's Oster Soy** reconstituted	97	Tr	1.07	0.50	2.7	0.04	0.05	0.7	0.03	0.1	3	0.23	1.1	7
62	**Formula S Soya Food**	600	0	8.70	10.00	40	0.30	0.80	3.0	0.30	1.6	80	2.0	12.0	60
63	**Formula S Soya Food** reconstituted	78	0	1.07	1.30	4.9	0.04	0.10	0.4	0.04	0.2	10	0.30	1.5	8
64	**Prosobee**	390	0	8.11	11.60	78	0.39	0.46	6.2	0.31	1.5	78	2.30	39.0	42
65	**Prosobee** reconstituted	49	0	1.02	1.46	9.7	0.05	0.06	0.8	0.04	0.2	10	0.29	4.9	5
66	**Wysoy**	810	90	9.75	8.80	103	0.88	1.07	5.0	0.68	2.1	64	2.55	42	80
67	**Wysoy** reconstituted	105	12	1.30	1.14	13.3	0.11	0.14	0.6	0.09	0.3	8	0.33	5.4	10
68	**Farley's Junior Milk**	570	(34)	7.80	3.40	19	0.29	1.10	4.6	0.29	1.4	50	2.60	21.0	46
69	**Farley's Junior Milk** reconstituted	78	(5)	1.07	0.47	2.6	0.04	0.15	0.6	0.04	0.2	7	0.35	2.9	6
70	**Progress**	860	87	9.30	9.40	56	1.23	1.80	4.7	0.65	1.0	80	2.49	18.7	98
71	**Progress** reconstituted	120	12	1.30	1.31	7.9	0.17	0.25	0.7	0.09	0.1	11	0.35	2.6	14

RECIPES

- Unless specified the recipes use whole pasteurised milk, fresh cream, Cheddar cheese, non-dairy vanilla ice cream and plain white flour.
- An egg has been assumed to weigh 50g. A level teaspoon refers to a standard 5 ml spoon and has been taken to hold 5g salt.

73-75   Bournvita made up with milk

2 heaped tsp Bournvita powder (8g)
200 ml whole, semi-skimmed or skimmed milk

Boil milk, add powder and mix.

77-79   Build-up made up with milk

1 sachet Build-up (38g)
284ml whole, semi-skimmed or skimmed milk

81   Cambridge Diet powder made up with water

1 sachet Cambridge Diet powder
(39g chocolate   37.5g other flavours)
230-250ml water

83-85   Cocoa made up with milk

4g cocoa powder   4g sugar
200ml whole, semi-skimmed or skimmed milk

Mix powder with a little cold milk. Add sugar. Pour on boiling milk and stir briskly.

87   Savoury Complan made up with water

1 sachet savoury Complan (57g)
200ml water

89   Sweet Complan made up with water

1 sachet sweet Complan (58g)
200ml water

90-92   Sweet Complan made up with milk

1 sachet sweet Complan (58g)
200ml whole, semi-skimmed or skimmed milk

94-96 Drinking chocolate made up with milk

18g drinking chocolate powder
200ml whole, semi-skimmed or skimmed milk

Boil milk, add powder and mix.

98-100 Horlicks made up with milk

25g Horlicks powder
200ml whole, semi-skimmed or skimmed milk

Boil milk, add powder and mix.

102 Microdiet made up with water

1 sachet micro drink powder (33g)
250ml water

105-107 Milk shake powder made up with milk

15g powder
200ml whole, semi-skimmed, or skimmed milk

Stir powder into milk.

109-111 Ovaltine made up with milk

25g Ovaltine powder
200ml whole, semi-skimmed or skimmed milk

Boil milk, add powder and mix.

199 Banana split

2 banana halves
37ml double cream
1 glacé cherry

120g ice cream
6g chopped nuts

Cover both halves of banana with ice cream. Top with whipped cream. Sprinkle with nuts, add cherry.

201 Chocolate nut sundae

115g ice cream
45ml double cream
70g chocolate sauce

6g chopped nuts
wafer (1g)

Cover ice cream with whipped cream and chocolate sauce. Sprinkle with nuts, add wafer.

210 Ice cream with cone

110g soft serve ice cream
1 cone (5g)

211 Ice cream with wafers

50g ice cream
2 ice cream wafers (2g)

213 Knickerbocker glory

90g jelly cubes
70g peach slices
70g pineapple chunks

160g ice cream
50ml double cream
2 glacé cherries

Prepare jelly as No. 237, allow to set and chop. Put fruit in the base of 2 sundae dishes, layer with jelly and ice cream. Top with whipped cream and a cherry.

214 Kulfi

600ml whole milk
450ml double cream
2g almond essence

3 tbsp honey
50g chopped almonds

Heat milk until it boils and continue boiling to reduce to 450 ml. Stir in honey and cream. Continue to stir while the volume reduces further by one third. Add almonds and almond essence. Pour into a container and freeze. When at freezing point, whip and re-freeze until solid.

Weight loss: 42.5%

215 Peach melba

113g ice cream
110g tinned peaches

50ml whipping cream
2g wafer

Cover two peach halves with ice cream. Top with whipped cream and a wafer.

216 Lemon sorbet

568ml water
224g caster sugar
grated rind of 2 lemons

140ml lemon juice
65g egg whites

Boil water, lemon rind and sugar for 10 mins. Leave to cool and strain. Strain lemon juice and add to syrup. Freeze until nearly firm then beat in stiff egg whites. Continue freezing until firm.

Weight loss: 32.5% for syrup

217 Blancmange

568ml milk
43g cornflour
43g sugar
essence and colouring as desired

Mix sugar and cornflour and dilute with a small quantity of cold milk. Boil remaining milk and whisk into cornflour mixture. Heat until mixture reboils and thickens, remove from heat and turn into moulds. Allow to set.

Weight loss: 7.5%

218 Cheesecake

Base for 18cm tin: 150g digestive biscuits
75g margarine

Top:
350g cream/curd cheese 2 eggs
25g cornflour 100g caster sugar
140ml double cream juice and rind of one lemon (70g)
½ tsp vanilla essence

Melt margarine in a pan and combine with biscuit crumbs. Press into base of tin. Combine topping ingredients and pour into base. Bake for 45 mins at 180°C/mark 4, until only just firm in the centre.

Weight loss: 5.6%

221 Creme caramel

3 eggs 113g sugar
284ml milk 70ml water
28g caster sugar vanilla essence

Prepare caramel by boiling sugar with about three quarters of the water until it caramelises. Add remaining water and reboil. Pour caramel into moulds. Beat eggs and milk, add caster sugar and vanilla essence and strain into moulds. Bake in a bain marie at 160°C/mark 3 for 45 mins to 1 hr.

Weight loss: 13.7%

222-224 Custard made up with milk

500ml whole, semi-skimmed or skimmed milk
25g custard powder
25g sugar

Blend custard powder with a little of the milk. Add sugar to remainder of milk and bring to the boil. Pour immediately over paste, stirring all the time. Return to pan, bring back to boiling point while stirring.

Weight loss: 20.9%

226 Confectioners' custard

1 egg 14g flour
28g caster sugar 150ml milk
2-3 drops vanilla essence

Cream sugar and egg, then fold in flour. Stir in milk. Either stir over a gentle heat until thickened or bake in a dish standing in a pan of water at 170°C/mark 3 until set.

Weight loss: 20.6%

227 **Egg custard**

500ml milk   2 eggs
30g sugar   vanilla essence

Beat the eggs and sugar together, add milk and vanilla essence. Either stir over a gentle heat until thickened or bake in a dish standing in a pan of water at 170°C/mark 3 for 40 mins.

Weight loss: 20.6%

229-231 **Dream Topping made up with milk**

1 packet Dream Topping powder (40g)
150ml whole, semi-skimmed or skimmed milk

Sprinkle powder onto milk. Whisk briskly.

232 **Fruit fool**

400g fruit   375ml water
35g cornflour   185ml whipping cream
100g sugar

Dilute cornflour with a little water. Purée fruit with the remaining water and sugar. Stir in cornflour and reboil. Allow to cool. Lightly whisk the cream and fold into the mixture.

Weight loss: 36.2% for fruit purée

234-236 **Instant dessert powder made up with milk**

1 packet instant dessert powder (66g)
300ml whole, semi-skimmed or skimmed milk

Sprinkle powder onto milk. Whisk briskly.

237 **Jelly made with water**

130g jelly cubes   440ml water

Dissolve jelly cubes in hot water. Add rest of the cold water. Pour into a mould and allow to set.

238-240 **Jelly made with milk**

130g jelly cubes   250ml milk
200ml water

Dissolve the jelly cubes in hot water. Cool, add milk slowly, stirring constantly. Leave to set in a mould.

241-243 **Milk puddings**

500ml whole, semi-skimmed or skimmed milk
50g rice, sago, semolina or tapioca
25g sugar

Simmer until cooked or bake in a moderate oven at 180°C/mark 4.

Weight loss: 19.1%

249 Trifle

75g sponge cake	250g custard
25g jam	25ml double cream
50g fruit juice	10g mixed nuts
75g tinned fruit	10g angelica and cherries
25ml sherry	

Slit sponge cake, spread with jam and sandwich together. Cut into 4cm cubes. Soak in the fruit juice and sherry. Mix with fruit, cover with cold custard and decorate with whipped cream, nuts and angelica.

251 Trifle with Dream Topping

As trifle (No. 249) except substitute 25g of made up Dream Topping for 25g double cream.

261-263 Bread sauce

250ml whole, semi-skimmed or skimmed milk	
50g fresh breadcrumbs	2 cloves
5g margarine	mace
½ tsp salt	1 small onion

Put milk and onion, stuck with cloves, in a saucepan and bring to the boil. Add breadcrumbs, and simmer for about 20 minutes over gentle heat. Remove onion, stir in margarine and season.

Weight loss: 6.8%

264 Cauliflower cheese

100g cheese, grated	25g margarine
1 small cauliflower (700g)	25g flour
100ml cauliflower water	250ml milk
½ level tsp salt	pepper

Boil cauliflower until just tender, break into florets. Drain saving 100ml water, place in a dish and keep warm. Make a white sauce from the margarine, flour, milk and cauliflower water. Add 75g cheese and season. Pour over the cauliflower and sprinkle with the remaining cheese. Brown under a grill or in a hot oven, 220°C/mark 7.

Weight loss: 14.6%

265 Cheese and potato pie

455g boiled potatoes	
30g margarine	140g milk
1g salt	55g egg
55g cheese	

Mash potatoes with margarine and salt. Beat egg with milk and mix into potatoes. Grate cheese, add half to the potato mixture. Turn into a dish and top with remaining cheese. Bake at 190°C/mark 5 for 45 mins to 1 hr.

Weight loss: 7%

266 **Cheese pastry**

60g cheese
140g shortcrust pastry

Proportions are derived from recipe review.

Weight loss: 12%

267 **Cheese pudding**

50g fresh breadcrumbs	cayenne pepper
250ml milk	75g grated cheese
½ level tsp salt	2 eggs

Heat milk, pour over breadcrumbs and soak for 30 mins. Add grated cheese, seasoning and egg yolks. Fold in stiffly whipped egg whites and pour into a greased pie dish. Bake at 180°C/mark 4 for 30 mins until well risen and golden brown.

Weight loss: 9.8%

268-270 **Cheese sauce**

350ml whole, semi-skimmed or skimmed milk
75g cheese	25g flour
½ level tsp salt	25g margarine
cayenne pepper	

Melt the fat in a pan, add flour and cook gently for a few minutes stirring all the time. Add milk and cook until mixture thickens, stirring continually. Add grated cheese and seasoning. Reheat to soften the cheese, serve immediately.

Weight loss: 15.2%

272-274 **Cheese sauce, packet mix, made up**

1 pkt cheese sauce mix (33g)
284ml whole, semi-skimmed or skimmed milk

Prepared as packet directions.

Weight loss: 9.1%

275 **Macaroni cheese**

280g cooked macaroni	25g flour
350ml milk	100g grated cheese
25g margarine	½ tsp salt

Boil macaroni and drain well. Make a white sauce from the margarine, flour and milk. Add 75g of the cheese and season. Add the macaroni and put in a pie dish. Sprinkle with remaining cheese and brown under grill or in a hot oven at 220°C/mark 7.

Weight loss: 9.4%

277 Mayonnaise

1 egg yolk	¼ level tsp made-up mustard
125g olive oil	20ml vinegar
¼ level tsp salt	pepper

Beat yolk and seasoning in a bowl. Whisk oil in very gradually to form a thick emulsion adding the vinegar slowly.

280-282 Onion sauce

350ml whole, semi-skimmed or skimmed milk

200g cooked onion	25g flour
1 level tsp salt	25g margarine
pepper	

Make the white sauce (as nos. 296-298), add the chopped onion and seasoning.

Weight loss: 12.6%

283-284 Cheese and egg quiche

200g raw plain or wholemeal shortcrust pastry

150g cheese	3 eggs
150g milk	

Line a 20cm flan ring with the shortcrust pastry. Fill with grated cheese. Beat eggs in the warmed milk and pour into pastry case. Bake for 10 mins at 200°C/mark 6 and then 30 mins at 180°C/mark 4.

Weight loss: 10%

285-286 Quiche Lorraine

200g raw plain or wholemeal shortcrust pastry

2 eggs	100g streaky bacon
200ml milk	100g cheese

Line a 20cm flan ring with shortcrust pastry. Fill with the fried, chopped bacon and grated cheese. Beat the eggs in warmed milk and pour into the pastry case. Bake for 10 mins at 200°C/mark 6, then for 30 mins at 180°C/mark 4.

Weight loss: 25.5%

287-288 Mushroom quiche

200g raw plain or wholemeal shortcrust pastry

100g sliced mushrooms	2 eggs
100g cheese	100ml milk

As Quiche Lorraine (Nos 285-286) except omit bacon and add mushrooms.

Weight loss: 10%

289 Raita

225g butter
454g low fat plain yogurt
80g mixed flavourings for curry

5g garlic
1050ml water

Bring ingredients to the boil and simmer gently.

Weight loss: 28.5%

290 Yogurt and gram flour raita

30g gram flour
300g low fat plain yogurt
300g onion
3g mixed spices
8g salt

300ml water
45g butter ghee
15g coriander leaves
3g turmeric

Make flour into a paste with some of the water, add yogurt and remaining water. Fry chopped onions in ghee, add coriander leaves, mixed spices, turmeric and salt and then flour and yogurt mixture. Keep stirring the mixture until it boils and thickens.

Weight loss: 25%

293 Tzatziki

250g Greek cows milk yogurt
5g fresh garlic
fresh chopped mint

213g cucumber
4g salt

Recipe from yogurt manufacturers.

294-295 Welsh rarebit

2 slices buttered toast (white or wholemeal bread)
50g grated cheese
¼ level tsp salt
pepper

20ml milk
¼ level tsp dry mustard
cayenne pepper

Mix cheese and seasoning with the milk. Spread on the toast. Brown under grill.

Weight loss: 8.1%

296-298 Savoury white sauce

350ml whole, semi-skimmed or skimmed milk
25g flour
½ level tsp salt

25g margarine

Melt fat in a pan. Add flour and cook for a few minutes, stirring constantly. Add milk and salt, and cook gently until mixture thickens.

Weight loss: 18.1%

299-301 Sweet white sauce

350ml whole, semi-skimmed or skimmed milk
25g flour
25g margarine

30g sugar

As savoury white sauce (Nos 296-298) except adding sugar and omitting salt.

Weight loss: 16.7%

303-305 White sauce packet mix, made up

1 pkt white sauce mix (22g)
284ml whole, semi-skimmed or skimmed milk

Prepared as packet directions.

Weight loss: 7.5%

811 Scrambled eggs with milk

2 eggs	20g butter
15ml milk	½ level tsp salt

Melt butter in pan, stir in beaten egg, milk and seasoning. Cook over gentle heat until mixture thickens.

Weight loss: 10.9%

817 Curried omelette/egg masala

200g egg	5g salt
125g butter	75g raw onion
20g curry powder	

Recipe proportions from survey work.

Weight loss: 18.6%

818 Egg fried rice

35g oil	1½ beaten eggs
45g chopped onion	350g cooked white rice

Heat oil, add egg and remaining ingredients. Cook, turning mixture over, for 3 mins.

Weight loss: 17.3%

819 Egg fu yung

45ml oil	50g chopped almonds
40g sliced spring onions	30ml sherry
100g beansprouts	15ml soya sauce
50g mushrooms	6 beaten eggs

Stir fry beansprouts, mushrooms, onions and almonds in most of the oil. Pour in sherry and soya sauce, bring to the boil then remove from heat. Add to beaten eggs. Heat remaining oil in pan, pour in ¼ mixture, stir and cook until brown, turn over and cook other side. Repeat process.

Weight loss: 17.2%

820 Egg nog

1 egg	20g sugar
284ml milk	50ml sherry/brandy

Whisk egg and sugar, add sherry/brandy. Heat milk and pour it over the egg mixture. Stir well.

821 Omelette

2 eggs
10ml water
10g butter

½ level tsp salt
pepper

Beat eggs with salt and water. Heat butter in an omelette pan. Pour in the mixture and stir until it begins to thicken evenly. While still creamy, fold the omelette and serve.

Weight loss: 5.7%

822 Cheese omelette

115g omelette, cooked
60g Cheddar cheese

Proportions are derived from recipe review.

823 Spanish omelette

4 eggs
90g chopped onion
60g diced boiled potato
130g tomato
154g green/red pepper

25g oil
5g garlic
salt
pepper
60g cooked peas

Fry onions in oil and add peppers, tomatoes, peas, garlic and potatoes and continue frying for 2-3 mins. Beat eggs, season, pour over vegetables. When set, fold and serve.

Weight loss: 19.4%

824-825 Scotch eggs

4 eggs
250g raw pork sausage meat
25g dried breadcrumbs

20g flour
15g beaten egg

Hard boil the eggs, cool and shell. Dip in seasoned flour, cover with sausage meat. Brush with beaten egg and coat with crumbs. Deep fry for 8-10 mins.

Weight loss: 2%

826 Plain soufflé

50g margarine
50g flour
4 eggs

1g salt
250g milk

Melt the margarine over gentle heat, stir in flour and add milk slowly. Bring to the boil and cook for minute or two. Cool slightly, beat in egg yolks and seasoning. Whisk egg whites until stiff and fold into mixture. Bake in a greased 17cm soufflé dish at 200°C/mark 6 for 35 mins.

Weight loss: 10%

827 **Cheese soufflé**

50g margarine
50g flour
4 eggs
½ level tsp cayenne pepper

½ level tsp salt
250ml milk
100g grated cheese
½ level tsp dry mustard

As plain soufflé (No 826). Add cheese with egg yolks and seasoning.

Weight loss: 15.2%

828 **Macaroons**

196g sugar
112g ground almonds
14g rice flour
3 egg whites

½ tsp vanilla essence
rice paper
chopped almonds

Beat egg whites until stiff. Fold in sugar, ground almonds, rice flour and vanilla essence. Pipe the mixture in 1 inch rounds onto the rice paper. Sprinkle with chopped almonds and bake for 20 mins at 180°C/mark 4.

Weight loss: 16%

829 **Meringue**

4 egg whites   200g caster sugar

Whisk egg whites until stiff. Fold in the sugar. Pipe onto baking sheet and bake at 130°C/mark ½ for 3 hrs.

Weight loss: 33.3%

830 **Meringues filled with cream**

40% meringue   60% whipping cream

Proportions derived from a number of shop-bought samples.

REFERENCES TO TABLES

1   Department of Health and Social Security (1977) *The composition of mature human milk. Report on Health and Social Subjects No. 12*, HMSO, London

2   Englyst, H. N. and Cummings, J. H. (1988) An improved method for the measurement of dietary fibre as the non-starch polysaccharides in plant foods. *J. Assoc. Off. Anal. Chem.* **71**, 808-814

3   Posati, L. P. and Orr, M. L. (1976) *Composition of foods, dairy and egg products, raw, processed and prepared*, Agriculture Handbook No. 8-1, US Department of Agriculture, Washington DC

4   Wenlock, R. W., Sivell, L. M., and Agater, I. B. (1985) Dietary fibre fractions in cereal and cereal-containing products in Britain *J. Sci. Food Agric.* **36**, 113-121

5   Wharton, P. A., Eaton, P. M. and Day, K. C. (1983) Sorrento Asian food tables: food tables, recipes and customs of mothers attending Sorrento Maternity Hospital, Birmingham, England. *Hum. Nutr.: Appl. Nutr.* **37A**, 378-402

6   Wiles, S. J., Nettleton, P. A., Black. A. E. and Paul, A. A. (1980) The nutrient composition of some cooked dishes eaten in Britain: A supplementary food composition table. *J. Hum. Nutr.* **34**, 189-223

FOOD INDEX

- Foods are indexed by their food number and **not** by page number.

- Numbers in brackets refer to foods where limited nutritional information is given in footnotes. Other aspects of the composition may differ from the values given in the main tables.